A Simple Guide to Using the
iPhone 12, Mini, Pro, & Pro Max

A Simple Guide to Using the

iPhone 12

Mini, Pro, & Pro max

A Complete User Manual for Beginners — with Useful
Tips and Tricks

Dylan Blake & Patrick Garner

While every precaution has been taken in preparing this book, the publisher does not claim any liability for mistakes or omissions, or damages arising from using the information contained herein.

A Simple Guide to Using the iPhone 12, Mini, Pro, & Pro max

First edition. November 24, 2020.

ISBN: 9798571175395

Written by Dylan Blake and Patrick Garner

Contents

INTRODUCTION

Congratulations on purchasing one of Apple's newest flagship phones. Now that you have this beauty, it's time to put to good use all it has to offer by making use of a well-illustrated user manual that unlocks all its hidden tips and tricks. This user guide has been specially constructed to give you precisely what you need, with so many tips and tricks to aid you to become familiar with the iPhone 12 Pro Max, iPhone 12 Pro, iPhone 12, and iPhone 12 mini. If you have an older iPhone model, you don't have to feel left out because this user guide was written with you in mind as well.

The iPhone 12 models come in different sizes. The iPhone 12 Pro Max measures in at 6.7-inches. While the iPhone 12 Pro and the iPhone 12 both features a 6.1-inch screen, similar to the 6.1-inch size of the iPhone 11. The iPhone 12 mini has no comparison when it comes to size. It's the world's smallest, thinnest, lightest 5G phone; it features a 5.4-inch display.

The iPhone 12 models are the first iPhone to feature support for 5G connectivity. There is support for mmWave 5G in the United States exclusively, and in other countries (including the U.S), there is support for sub-6GHz 5G. Although there is no killer app for 5G right now, it will be a game-changer in the nearest future. So, iPhone 12's 5G support will ultimately future-proof your phone.

The iPhone 12 models house some mouth-watering features. It boasts the new, faster A14 Bionic chip, which delivers a massive boost in performance. They are the first phone devices capable of real-time Dolby vision recording (a professional HDR video format). They are the first iPhone to feature Super Retina XDR OLED displays with a Ceramic Shield cover for improved durability and 4x better drop protection. They also feature MagSafe wireless chargers (a magnetic wireless charging system), which is two times faster than the older Qi wireless charger. A ring of magnets encircles the wireless charging coil of the iPhone 12, making it possible for MagSafe to snap right in place at the back of your iPhone. MagSafe is sold separately.

As usual, the main difference between the iPhone 12 models and the iPhone 12 Pro models is the camera. While the iPhone 12 Pro models include a triple-lens camera setup with a LiDAR Scanner and other bells and whistles, iPhone 12 and iPhone 12 mini houses a more straightforward and less advanced dual-lens camera setup,

Section 1: Getting Started

1: Unboxing and Setting up the iPhone 12

The iPhone 12 models come in a thinner box (relatively to its predecessors) because Apple didn't give room for power adapter or EarPods. Apple stated that they didn't include these accessories as part of their plans to reach their environmental goals. You can use your current Apple power adapter and headphones or purchase these items separately. Apart from this arbitrary omission, the iPhone 12 is still a great phone, worth every penny.

The other thing in your iPhone 12 box is a USB-C to Lightning Cable, and it is imperative to note that this is not your typical charging cable, so it might not fit into your computer USB port or the wall outlets in airports, cafes, or similar places. Below is a list of things to purchase for a better experience.

1. Apple 20W USB-C power adapter
2. iPhone 12 headphones
3. MagSafe: iPhone 12 wireless charger
4. iPhone 12 smartwatch

Power on and set up iPhone

An internet connection is needed to power on and set up your iPhone. You can also use a computer to set up your iPhone. You can transfer your data to your new iPhone if you have an old iPhone, iPad, iPod touch, or an Android device.

Get ready for setup

Get everything you'd need to make the setup as smooth as possible. Have the following items available:

- A cellular data service through a carrier or an internet connection through a Wi-Fi network
- Your Apple ID and password; you can create one during setup if you don't have an Apple ID.
- Your debit or credit card account information, if you would like to add a card to Apple Pay during setup.
- Your previous iPhone or Android device, if you are transferring your data to your new device.

Power on and set up your iPhone

1. Press and hold the side button or Wake or Sleep button (depending on your iPhone model) until you see the Apple logo.

Side button   Side button

Sleep or Wake button

2. Do one of the following:

    - Tap **Set Up Manually**, then follow the
      prompts.
    - You can use Quick Start to automatically set
      up your device if you have an additional
      iPhone, iPad, or iPod touch with iOS 11,
      iPadOS 13, or later. Bring the two devices close
      to each other, then follow the prompts to
      securely copy many of your settings,
      preferences, and iCloud Keychain. You can
      then use your iCloud backup to restore the
      rest of your data and content to your new
      device.

Or, you can transfer all your data wirelessly from your previous device to your new one if the two devices have iOS 12.4, iPadOS 13, or later.

Set up your cellular service

A SIM from a carrier is required to set up the cellular connection on your iPhone; contact your carrier to set up a cellular plan.

The iPhone 12 models can connect to 5G networks. All you need is any iPhone 12 models, a carrier that supports 5G, and a 5G cellular plan.

The iPhone XS, iPhone XR, and later (to iPhone 12) all support Dual SIM using one physical nano-SIM and eSIM (not available in all regions or countries).

What is an eSIM?

It is a recently developed technology that allows you to easily switch between carriers and have more than one phone number on your mobile phone. An eSIM, which stands for "embedded subscriber identity module," is a small electronic chip implanted into a mobile device to serve the same function as the little plastic SIM cards.

Below are some of the numerous ways you can use Dual SIM:

- You can use one number for voice plans and the other for data plans.
- You can use one number for personal calls and the other for business calls.
- You can add a local data plan when you travel to another country or region.

Install a physical nano-SIM

1. Insert a SIM eject tool or a paper clip into the small hole on the SIM tray, then push it in to eject the tray

Sim tray

Paper clip or Sim eject tool

2. Remove the tray from iPhone and place the nano-SIM in the tray (the angle corners determines the right orientation)

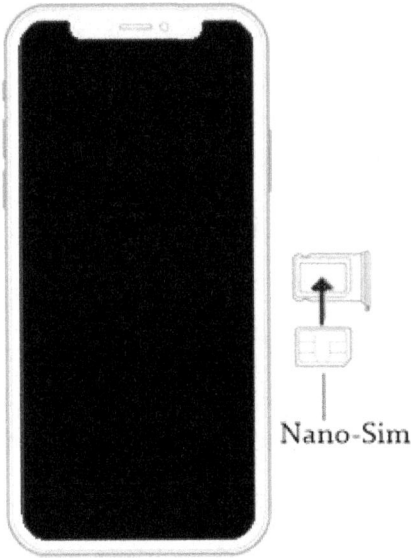

Nano-Sim

3. Then insert the SIM tray back into the iPhone.

Set up an eSIM cellular plan

On iPhone XS, iPhone XR, and later (to iPhone 12), you can digitally store an eSIM provided by your carrier.

1. To do so, go to **Settings** ⚙ > **Cellular**, and tap **Add Cellular Plan**.

2. Then do one of the following:

   • **Use a QR code provided by your carrier to set up a new plan**: Position your iPhone camera so that the QR code appears in the frame, or manually enter the details. You may be asked to enter your carrier's confirmation code,

- **Install an assigned cellular plan**: If your carrier sent you a notification that a plan was assigned to you, click on **Carrier Cellular Plan Ready to Be Installed.**

- **Transfer a SIM from your old iPhone to your new iPhone**: Choose your phone number from the list. If you can't find your phone number in the list, make sure you are signed in to the same Apple ID on both iPhone devices.

3. Click on **Add Cellular Plan.**

4. In cases where the new plan is your second line, follow the prompts to set how you want the plans to work together.

Manage your cellular plans for Dual SIM

During setup on iPhone models with Dual SIM (iPhone XS, iPhone XR, and later), you can choose how you want the iPhone to use each line. If you want to adjust the setting later, do any of the following:

1. Go to **Settings** ⚙ > **Cellular**, and do the following:

- Click on **Cellular Data**, and select a default line. Turn on **Allow Cellular Data Switching** to use either line depending on availability and coverage.
- Tap **Default Voice Line**, and choose a line.
- Tap a line below Cellular Plans if you want to change settings such as Cellular Plan Label, Wi-Fi Calling, Calls on Other Devices, or SIM PIN.

Connect your iPhone to the internet

Use Wi-Fi or a cellular network to connect your iPhone to the internet.

Moving from an Android device to iPhone

You can securely and automatically move your data from an Android device when you first set up your new iPhone.

N.B. It is only when you first set up the iPhone that you can move to the iOS app. You must erase your iPhone and start over, or move your data manually if you've already finished setup and want to use Move to iOS.

1. Do the following on your iPhone:
   - Follow the setup assistant.

- Then tap **Move Data from Android** on the Apps & Data screen.

2. Do the following on the Android device:

- Turn on Wi-Fi.
- Open the Move to iOS app.
- Then follow the prompts.

Connect your iPhone to a Wi-Fi network

1. Go to **Settings** ◉ > **Wi-Fi**, then turn on Wi-Fi.

2. Do either one of the following:

- **A network**: Enter the password (if required).
- **Other**: Tap this to join a hidden network. Then enter the hidden network's name, security type, and password.

3. Your iPhone is connected to a Wi-Fi network when

🛜 appears at the top of the screen (to verify this, open Safari or any web browser to view a webpage)

Connect your iPhone to a personal Hotspot

You can connect to the cellular network of a friend, family, or colleague when they share their Hotspot.

1. Go to **Settings** ◉ > **Wi-Fi**, then select the name of the device sharing the Personal Hotspot.

2. If you asked for a password, enter the password shown in **Settings > Cellular > Personal Hotspot** on the device sharing the Personal Hotspot.

Connect your iPhone to a cellular network

If a Wi-Fi network isn't available, your iPhone automatically connects to your carrier's cellular data network.

Manage Your Apple ID and iCloud setting on iPhone

Your Apple ID account is needed to access Apple services, e.g., the iTunes Store, App Store, iCloud, iMessage, etc. iCloud helps you securely store your documents, photos, videos, music, apps, etc., across all your Apple devices. You can also use iCloud to easily share photos, calendars, locations, etc., with friends and family.

Sign in with your Apple ID

Do the following if you didn't sign in during setup:

- Go to **Settings** ⚙, **then tap Sign in to your iPhone**.
- Enter your Apple ID and password, or create one if you don't have an Apple ID.
- Enter the six-digit verification code if you protect your account with two-factor authentication.

Change your Apple ID settings

1. Go to **Settings** ⚙ > [your name], then do any of the
    following:

    -   Update your contact information.
    -   Change your password.
    -   Manage Family Sharing.

Change your iCloud settings

1. Go to **Settings** ⚙ > [your name] > **iCloud**, then do
    any of the following:

    -   See your iCloud storage status.
    -   To Upgrade your iCloud storage, tap **Manage
        Storage** > **Change Storage Plan**.
    -   Turn on the apps that should use iCloud, such
        as Messages, Photos, Contacts, Calendar, etc.

Turn on the app
that should use iCloud

2: The Nitty-gritty

Before we go into more advanced ways of operating your new iPhone device, we'll be looking at some basic features that will help you efficiently manage, control, and navigate your phone.

Wake and unlock your iPhone

By default, the iPhone automatically turns off the display and goes to sleep to save power and prevent unauthorized access to your iPhone.

Wake your iPhone

Do one of the following to wake your iPhone:

- Press and hold the side button or Wake/Sleep button (depending on your iPhone model) until you see the Apple logo.

Side button   Side button

Sleep or Wake button

- Raise your iPhone (if you want to turn on or off this feature, go to **Settings** ◉ > **Display & Brightness**).

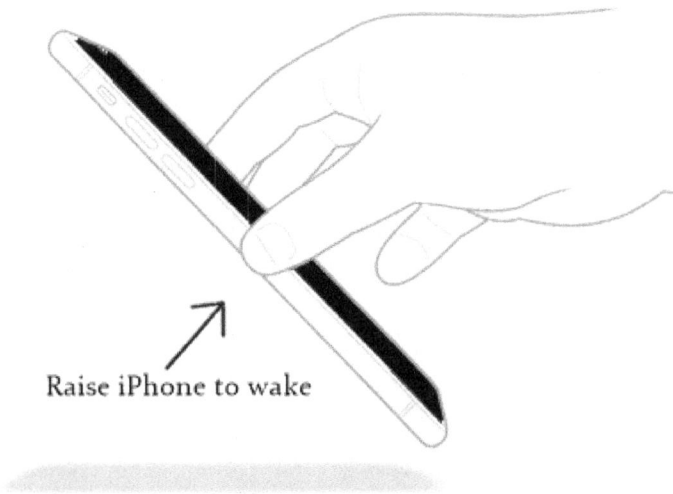

Raise iPhone to wake

- Tap the screen (available on iPhone X, iPhone XS, iPhone XR, and later).

Use Face ID to unlock your iPhone

If you didn't set up Face ID when you set up your iPhone, go to **Settings** ⚙ > **Face ID & Passcode** > **Set Up Face ID**, then follow the prompts (available on iPhone X, iPhone XS, iPhone XR, and later).

- Raise your iPhone or tap the screen to wake it, then glance at the screen. A lock icon animates from closed to open to indicate your iPhone is unlocked.
- Swipe up from the bottom of the screen.

Use Touch ID to unlock your iPhone

If you didn't set up Touch ID when you first set up your iPhone, go to **Settings** ⚙ and click on **Touch ID & Passcode**, turn on any of the options and follow the prompts (available on 1st generation iPhone SE, iPhone 6s Plus, to 2nd generation iPhone SE).

- Press the Home button with the finger you registered with Touch ID.

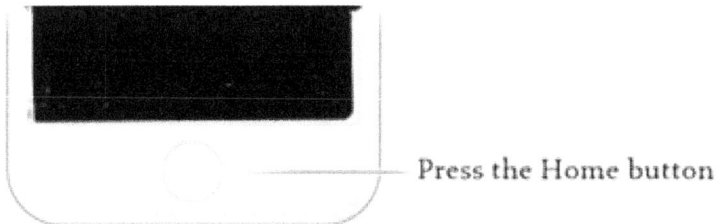

Press the Home button

Use a passcode to unlock your iPhone

If you didn't set up passcode when you set up iPhone, go to

Settings ⚙, tap **Face ID & Passcode** or **Touch ID &**

Passcode (depending on your iPhone model), then tap

Turn Passcode On.

After setting a passcode, you can use Face ID or Touch ID to

unlock your iPhone (depending on your model).

Gestures and tricks for iPhone models with Face ID

Learn important gestures and tricks to navigate your iPhone
easily.

- **Go Home**: Quickly return to the Home screen at
  any time by swiping up from the bottom edge of
  the screen.
  Gesture:

Swipe up

- **Open App Switcher**: Quickly open the App Switcher by swiping up from the bottom edge, then pause in the middle of the screen and lift your finger. Swipe right to browse the open apps, then tap the app you want to see.
  Gesture:

- **Access Control Center**: Quickly access the Control Center by swiping down from the top right corner.
  Gesture:

  Swipe down
  from the top right corner

- **Switch between open apps**: Quickly switch between open apps by swiping left or right along the bottom edge of the screen.

Gesture:

Swipe left or right
along the bottom edge

- **Take a screenshot**: Quickly take a screenshot of your current screen by simultaneously pressing and quickly release the volume up button and side button.

Gesture:

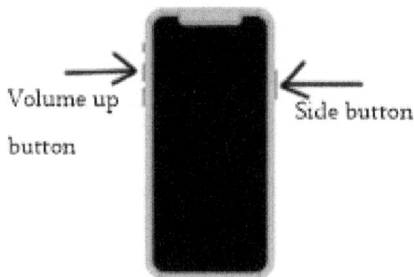

Volume up
button

Side button

- **Ask Siri**: Quickly talk to Siri by holding down the side button. Siri listens to you until you release the button. *See* Learn how to talk to Siri (page 219).

Gestures:

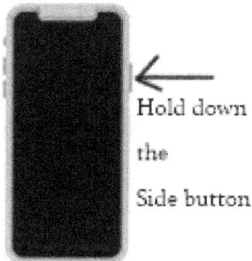

Hold down
the
Side button

- **Use Accessibility Shortcut**: Quickly access the Accessibility Shortcut by triple-clicking the side button.

  Gesture:

Triple-click
the
Side button

- **Use Apple Pay**: Quickly access Apple Pay by double-clicking the side button.

  Gesture:

Double-click
the
Side button

- **Use Emergency SOS**: Quickly access Emergency SOS by simultaneously pressing and holding the side button and either volume button until you see the sliders, then drag Emergency SOS.

  Gesture:

  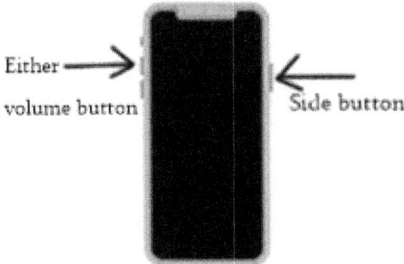

  Either → volume button   ← Side button

- **Power off**: Quickly power off your iPhone by simultaneously pressing and holding the side button and either volume button until you see the sliders, then drag the top slider to power off. Or go to **Settings** > **General** > **Shut Down**.

  Gesture:

  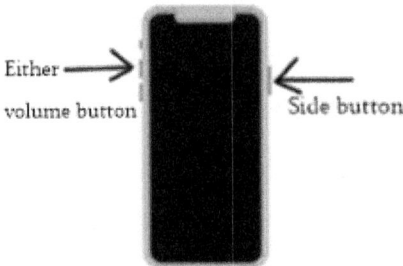

  Either → volume button   ← Side button

- **Force restart**: Quickly force-restart your iPhone by pressing and releasing the volume up button, pressing and releasing the volume down button, then press and hold the side button until you see the Apple logo.

Gesture:

Adjust your iPhone audio volume

You can use the buttons on your iPhone's side to adjust the audio volume when listening to music, on the phone, watching a movie, or other media. You can also use the buttons to control the volume for alerts, ringer, and other sound effects.

🎙 **Ask Siri**. You can say something like: "Turn down the volume" or "Turn up the volume." *See* Learn how to talk to Siri (page 219).

Volume up
button ——

Volume ——
down button

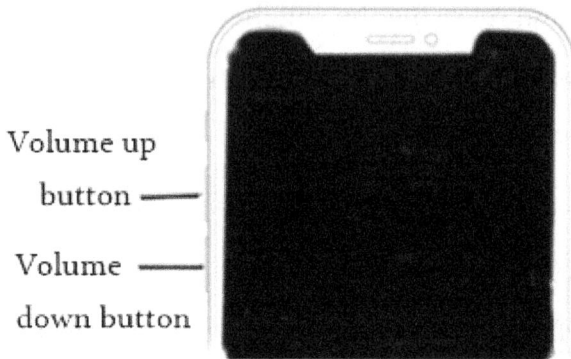

Put your iPhone in ring or silent mode

You can put your iPhone in ring mode △ or silent mode △ by flipping the Ring/Silent switch.

Flip the ——
Ring/Silent
switch

In ring mode, the iPhone plays all sound. In silent mode (the switch shows orange), iPhone doesn't ring or play alerts, notifications, and other sound effects.

N.B. Even when the iPhone is in silent mode, Clock alarms, Music, and games still play sounds via the built-in speaker.

Silence calls, alerts, and notifications temporarily

- Open Control Center by swiping down from the
  top-right corner of the screen (on iPhone with Face
  ID) or by swiping up from the bottom edge of the
  screen (on iPhone with a Home button), then tap

  🌙.

Change your iPhone alert tones and vibrations

You can change the sounds, and vibrations iPhone plays
when you get a call, text, email, voicemail, reminder, etc.

On iPhone 7, iPhone 7 plus, to later (iPhone 12), you feel a
tap (called haptic feedback) after performing specific
actions, e.g., when you touch and hold the Camera icon on
the Home Screen.

Set your alert tones and vibrations

1. Go to **Settings** ⚙ > **Sound & Haptics** (on iPhone 7,
   iPhone 7 plus, to later) or **Settings** ⚙ > **Sound** (on
   other iPhone models)

2. Drag the slider below Ringers and Alerts to set the
   volume for all sounds.

3. Tap a sound type, such as ringtone or text tone, to set
   the tones and vibration patterns for sounds.

4. You have the following options:

- **Choose a tone** (scroll to see all)

N.B. Ringtones are used for incoming calls, clock timer, and the clock alarm, while text tones are used for text messages, new voicemail, and other alerts.

- Tap **Vibration**, then select a vibration pattern, or tap **Create New Vibration** to create your vibration pattern.

Turn Haptic feedback off or on

- On iPhone 7, iPhone 7 plus, to later (iPhone 12), go to **Settings** ⚙ > **Sound & Haptics**
- Then turn System Haptics off or on
  N.B. You won't hear or feel vibrations for incoming calls and alerts when System Haptics is turned off.

Find settings on your iPhone

In the Settings app ⚙, you can search for settings that you want to change or adjust, e.g., notification sounds, passcode, etc.

- Tap Settings ⚙ on the Home Screen or in the App Library.

- To reveal the search field, swipe down, then enter a term ("Photos," for example).

Magnify your iPhone screen with Display Zoom

Use Display Zoom to see larger onscreen controls on your iPhone.

- Go to **Settings** ⚙ > **Display & Brightness**, then tap **View** below Display Zoom.
- Choose **Zoomed** > **Set**.

Change iPhone's name

You can change the name of your iPhone, which is used by Airdrop, iCloud, your Personal Hotspot, and your computer.

- Go to **Settings** ⚙ > **General** > **About** > **Name**.
- Tap ⊗, then enter a new name > **Done**.

Adjust your iPhone screen brightness and color

On iPhone, you can set Dark Mode, use Night Shift, and automatically adjust the screen base on lighting conditions.

Automatically adjust your screen brightness

iPhone uses the built-in ambient light sensor to adjust your screen brightness depending on the current light conditions.

- Go to **Settings** ◉ > **Accessibility**.
- Click on **Display & Text Size**, then turn on **Auto-Brightness**.

Manually adjust your screen brightness

Do either one of the following to make your iPhone dimmer or brighter:

- Open Control Center by swiping down from the top-right corner of the screen (on iPhone with Face ID) or by swiping up from the bottom edge of the screen (on iPhone with a Home button), then drag ☀.
- Go to **Settings** ◉ > **Display & Brightness**, and drag the slider.

Turn Dark Mode on or off

Utilize this feature and switch your iPhone to a dark color scheme that is perfect for low-light locations. You can turn on Dark mode
from Control Center or set it to automatically turn on at night (or on a custom schedule).

Switch your iPhone
to a dark color scheme
with Dark Mode

Turn Dark Mode on or off by doing either one of the
following:

- Open Control Center by swiping down from the top-
  right corner of the screen (on iPhone with Face ID) or
  by swiping up from the bottom edge of the screen (on
  iPhone with a Home button), touch and hold ☀,
  then tap ◐ to turn Dark Mode on or off.
- Go to **Settings** ⚙ > **Display & Brightness**, then
  select **Dark** to turn on Dark Mode, or select **Light** to
  turn it off.

Schedule Dark Mode to automatically turn on and off

- Go to **Settings** ⚙ > **Display & Brightness**.

- **Turn on Automatic > Options**.
- Select either Sunset to Sunrise or Custom schedule.

Schedule Night Shift to automatically turn on and off

Night Shift and Dark mode serve the same function; you can use the one that works best for you. The key difference between them is, while night mode changes the color being emitted by the screen to a warmer color, Dark Mode switches the user interface background to a darker shade along with any corresponding color scheme flips.

Use Night Shift to switch the colors in your display to a warmer end of the spectrum at night to keep your eyes more comfortable.

- Go to **Settings** ⚙ > **Display & Brightness** > **Night Shift**.
- Turn on Scheduled, then drag the slider below Color Temperature towards the warmer end of the spectrum to adjust the color balance for Night Shift.
- Tap **From**, then select either Sunset to Sunrise or Custom schedule.

Manually turn Night Shift on and off

This is helpful when you are in a darkened room during the day.

- Open Control Center by swiping down from the top-
  right corner of the screen (on iPhone with Face ID) or
  by swiping up from the bottom edge of the screen (on
  iPhone with a Home button), touch and hold ☀,
  then tap ☀ to turn Night Shift on and off.

Turn True tone on and off

On iPhone 8, iPhone 8 plus to later (iPhone 12), you can
switch on True Tone to automatically adjust the intensity
and color of the display to blend with the light in your
environment.

Do either one of the following:

- Open Control Center by swiping down from the top-
  right corner of the screen (on iPhone with Face ID) or
  by swiping up from the bottom edge of the screen (on
  iPhone with a Home button), touch and hold ☀,
  then tap ☀ to turn True Tone on and off.
- Go to **Settings** ⚙ > **Display & Brightness**, then turn
  True Tone on and off.

Open apps on iPhone

You can find and open your apps in the Home Screen, which
shows all your apps organized into pages.

You can also open your apps in the App Library, which organizes your apps in a simple, easy-to-navigate view.

Open your apps on the Home Screen

- To get to your Home Screen, swipe up from the bottom edge of the screen (on iPhone models with Face ID) or press the Home button (on iPhone models with a Home button).

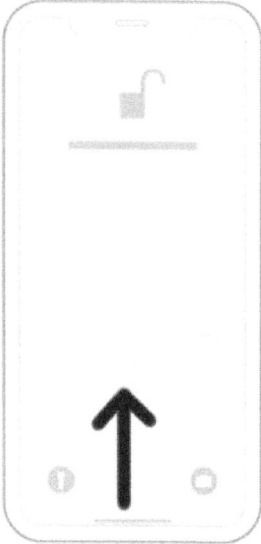

Swipe up from the bottom edge

- Browse apps on other Home Screen pages by swiping left or right.

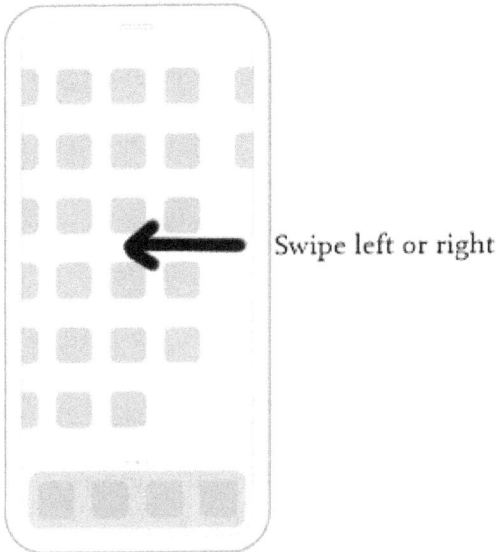

Swipe left or right

- Then tap the app's icon to open it.
- If you want to return to the first Home Screen page, swipe up from the bottom edge of the screen (on iPhone models with Face ID) or press the Home button (on iPhone models with a Home button).

Open your apps on the App Library

The App Library (i.e., a space at the end of your Home Screen pages) helps you quickly locate and open an app by automatically sorting your apps by category, e.g., Entertainment, Creativity, Social, etc. It also arranges the apps you use most near the top of the screen and at the top level of their categories.

- To go to your App Library, go to the Home Screen, then swipe left past all your Home Screen pages.

The App Library arranges your apps in categories.

- Then tap the app if it's visible.

Switch between your apps

Quickly switch from one app to another on your iPhone with the App Switcher. You can pick up right where you left off when you switch back.

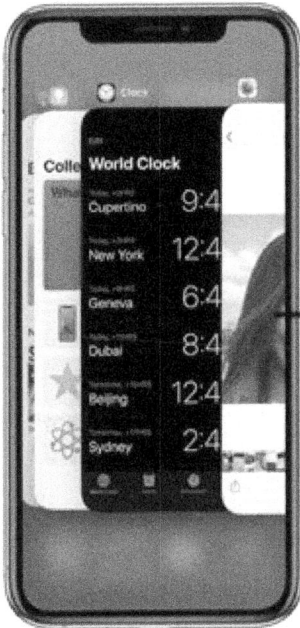

Quickly switch from one app to another with the App Switcher

Use the App Switcher

1. Do one of the following to see all your opened apps in the App Switcher:

    - Swipe up from the bottom edge of the screen and pause in the center (on iPhone models with Face ID). *See* Gestures and tricks for iPhone models with Face ID (page 16).

    - Double-click the Home button (on iPhone model with a Home button).

Switch between your opened apps

You can quickly switch between opened apps on iPhone models with Face ID by swiping left or right along the bottom edge of the screen. *See* Gestures and tricks for iPhone models with Face ID (page 16).

Organize your apps in folders

You can quickly and easily find apps on your Home Screen by grouping your apps in folders.

- Touch and hold any app on your Home Screen > **Edit Home Screen**.
- To create a folder, drag an app onto another app.
- To rename that folder, tap the name field, then input the new name.
- Drag all the apps out of the folder to automatically delete them.
- When you are finished, tap **Done**.

Quit an app

Open the App Switcher, swipe right to find the app, then swipe up on the app to remove it from open apps.

Delete an app on iPhone

Do any of the following to remove or delete an app from your iPhone:

- **Delete or remove an app from the Home Screen**: Touch and hold the app on the Home Screen to access the quick actions menu, tap **Remove App,** then tap **Move to App Library** if you want to keep it in the App Library, or tap **Delete App** to delete it from your iPhone.

- **Delete an app from the App Library and Home Screen**: Touch and hold an app on the App Library to access the quick actions menu, tap **Delete App > Delete**.

- **Add an app to the Home Screen in the App Library**: Touch and hold the app on the App Library to access the quick actions menu, then tap **Add to Home Screen**.

Take a screenshot or screen recording on your iPhone

You can capture the current image and record your screen's activities to share with family and friends or use in documents.

Take a Screenshot

1. To take a screenshot, do either one of the following:

   - **On iPhone with Face ID**: Simultaneously press and then release the volume up button and side button. *See* Gestures and tricks for iPhone models with Face ID (page 16).

   - **On iPhone with a Home button**: Simultaneously press and then release the Home button and the Sleep/Wake button or side button (depending on the iPhone model).

2. Tap the screenshot in the lower-left corner > **Done**.

3. Then choose your screenshot location.

Use the screen recorder

On your iPhone, you can create a screen recording and capture sound.

- Go to **Settings** ⚙ > **Control center**, then tap ⊕ behind Screen Recording.

- Open Control Center, tap ◉. Your screen recording starts after the three-second countdown.

- To stop recording, click on the red status bar at the top of the screen > **Stop**, or open Control Center, then tap ◉ .

Lock and unlock screen orientation on your iPhone

Many apps have their unique or different view when you rotate your iPhone. You can lock the screen orientation on your iPhone, so it doesn't change when you rotate it.

Rotate your iPhone

- To lock the screen orientation, open Control Center, then tap ⟲.

- ⟲ appears in the status bar when the screen orientation is locked.

Change your iPhone wallpaper

On iPhone, you can change the look and feel of your Lock and Home screens by picking your favorite pictures, pre-loaded wallpapers, or Live Photos.

Change iPhone wallpaper

Set your wallpaper

1. Go to **Settings** ⚙ > **Wallpaper** > **Choose a New Wallpaper**.

2. Do any of the following:

   - Choose a preset image from a group at the top of your screen, which consists of Dynamic, Stills, etc.

- Use one of your photos as wallpaper. Tap an album, then tap the photo.

- On some wallpaper choices, you can tap ⬚ to turn on Perspective Zoom, which makes your wallpaper seem to "be in motion" when you change your viewing angle.

3. Tap **Set**, then select one of the following:

   - **Set Lock Screen** (to set it as the wallpaper for your Lock screen).

   - **Set Home Screen** (to set it as the wallpaper for your Home Screen).

   - **Set Both** (to set it as the wallpaper for both your Lock Screen and Home Screen).

Use a Live Photo as your wallpaper for the Lock Screen

If you use a Live Photo as your wallpaper for the Lock Screen, touch and hold the Lock Screen to play the Live Photo.

1. Go to **Settings** ⚙ > **Wallpaper** > **Choose a New Wallpaper**.

2. Do either one of the following:

   - Tap **Live**, then choose a preset Live Photo.

   - Tap your Live Photo album, then choose a Live Photo.

3. Tap **Set**, then choose either Set Lock Screen or Set Both.

Search for apps

If you're finding it challenging to find an app or a setting, you can conveniently use the search feature to locate that app or setting.

Tap the search field

1. To access the search field, swipe down from the middle of the Home Screen.
2. Tap the search field, then input what you are searching for.

3. Do any of the following:

   - Tap **Go** on the keyboard to hide the keyboard
     and see more results on the screen.

   - Tap a suggested app to open it.

   - Tap $\otimes$ in the search field to start a new
     search.

Send items to nearby devices with AirDrop

Use AirDrop to send your pictures, videos, locations,
websites, etc., wirelessly to other nearby Apple devices and
MAC computers. To use AirDrop, Bluetooth and Wi-Fi must
be turned on. You also need to be signed in with your Apple
ID to use AirDrop.

Send an item to a nearby device

 - Open the item you want to send, then tap ⬆️, Share,
   AirDrop, ••• .

 - In the row of share options, tap ◉, then tap the
   profile picture of a nearby AirDrop user.

◌ **TIP:** On iPhone 11 and iPhone 12 models, position your
iPhone in the direction of another iPhone 11 and iPhone 12
model, then at the top of your screen, tap the profile picture
of its user.

Allow AirDrop to receive items

If you want to send an item to a person, but the person doesn't appear as a nearby AirDrop user, or you are the person on the receiving end, do this:

- Open Control Center, tap ⊚, then choose who you want to receive items from—tap **Contacts Only** or **Everyone**. You can decline or accept each request as it arrives.

Touch and hold the top-left group of controls if you don't see ⊚.

Use and Customize Control Center

On iPhone, Control Center gives you instant access to useful controls.

Open Control Center

- Swipe down from the top-right corner of the screen (on iPhone with Face ID). To close it, swipe up from the bottom

- Swipe up from the bottom edge of the screen (on iPhone with a Home button). To close it, press the Home button or swipe down.

Access other controls in Control Center

You can access more controls that offer additional options. Touch and hold a control to see available options. For instance, you can do the following in Control Center:

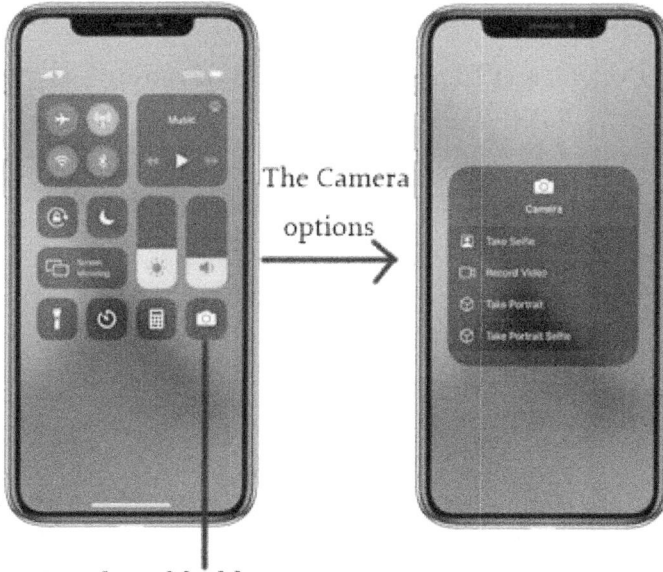

The Camera options

Touch and hold to see
Camera options

- Touch and hold the top-left group of controls, then to Open AirDrop, tap ⊙.

- Touch and hold ⬚ to see Camera options, e.g., Take Selfie, Record Video, etc.

Customize Control Center

You can customize Control Center by rearranging the controls and adding more controls and shortcuts to many apps, e.g., Notes, Calculator, etc.

1. Go to **Settings** ⚙ > **Control Center**.

2. Tap ⊕ or ⊖ next to a control to add or remove it.

3. To rearrange the controls, touch and hold ☰ next to a control, then drag it to a different position.

Widget

Today View widgets show you currrent data from your favorite apps at a glance

You can add widgets to your Home screen to get easy access to information or apps. Today View widgets show you current data from your favorite apps at a glance (that is, weather, today's headlines, calendar events, etc.)

Open Today View

To open this cool feature, from the left edge of the Home Screen or Lock Screen, swipe right.

Transfer a widget from Today View to the Home Screen

- Open Today View, touch and hold the widget you want to transfer until it starts to jiggle before dragging it off the right side of the screen.
- Drag the widget and place it where you want it on the Home Screen > **Done**.

Add a widget to a Home Screen page

- Go to the Home Screen page where you would like to add the widget, then touch and hold the Home Screen background until it begins to jiggle.
- To open the widget gallery, tap + at the top of the screen.
- Search or scroll to locate the widget you want, tap it, then swipe left through the size options.
- When you find the size you want, tap **Add Widget** > **Done**.

Remove a widget from the Home Screen

- To open the quick actions menu, touch and hold the widget.

- Tap **Remove Widget** (or **Remove Stack**) > **Remove**.

Allow access to Today view when your iPhone is locked

- Go to **Settings** ⚙ > **Face ID & Passcode** (on iPhone models with Face Id) or **Touch ID & Passcode** (on other iPhone models).

- Input your passcode, then below Allow Access When Locked, turn on **Today View**.

Section 2: Apps

The apps listed below are found on your iPhone. The apps are either preloaded or downloaded over-the-air (OTA) to your mobile Device during the setup.

Camera| Photos| Phone| Contacts| Messages| Calendar| Clock| Calculator| iPhone Health|

3: Camera

The camera of your iPhone 12 is remarkable. The details and clarity of images you can get from them are baffling. It is arguably the best camera in the cell phone industry right now.

Navigating the camera's photo screen

Photo mode is the standard mode that you see when you open Camera. Take better and cooler pictures by mastering how to navigate your camera's photo screen. Use Photo mode to take still photos. Swipe right or left to select a different mode, e.g., Video, Pano, Time-Lapse, Slo-Mo, and Portrait.

Photo mode is the standard mode
that you see when you
open Camera

Taking photos with your iPhone camera

- To get to your camera, tap 📷 on the Home screen or
  swipe left from the Lock screen to open the Camera
  in Photo mode.

- Press either volume buttons or tap the Shutter button
  to take the picture.

🎙 **Ask Siri**. Say something like: "Open Camera." *See* Learn
how to talk to Siri (page 219).

Turning the flash on or off

- On iPhone XS, iPhone XR, and later (to iPhone 12), tap ⚡ to switch the flash on or off. Or tap ⌃, then below the frame, tap ⚡ to choose Auto, On, or Off.

- While on iPhone X and earlier, tap ⚡, then choose Auto, On, or Off.

Zooming in or out

- On all models, pinch your camera screen to zoom in or out.

- On iPhone models with Dual (e.g., iPhone 12 and iPhone 12 mini) and Triple (e.g., iPhone 12 Pro and iPhone 12 Pro Max) camera systems, toggle between 1x, 2x, 2.5x, and .5x to zoom in and out quickly (depending on your model). To zoom in or out more efficiently, touch and hold the zoom controls, then drag the slider left or right.

Setting a timer

- On iPhone XS, iPhone XR, and later (to iPhone 12), tap ⌃, then tap ⏱.

- While on iPhone X and earlier, tap ⏱.

Taking a selfie

You can take cool selfies with the front camera, in Photo mode or Portrait mode (on iPhone X and later).

- Tap ◉ or ⬚ (depending on your model) to switch to the front camera.
- Hold your iPhone and position the front camera in front of you.

◉ **TIP:** You can tap the arrows inside the frame to increase the field of view (available only on iPhone 12 models and iPhone 11 models).

- Press either volume buttons or tap the Shutter button to take the picture.

Enable mirrored selfie to capture the picture exactly as you see it in the camera frame. Go to **Settings** ◉ > **Camera**, and turn on **Mirror Front Camera** (available on iPhone XS, iPhone XR, and later).

Adjusting the camera's focus and exposure

Before taking a picture, the iPhone camera automatically sets the focus and exposure (face detection balances the exposure across multiple faces). You can manually alter the focus and exposure by doing the following:

- Tap the screen to see the automatic focus area and exposure settings.

- Tap where you want to move the focus area to.

- Then drag ☀ next to the focus area up or down to adjust the exposure.

You can lock your manual focus and exposure settings for upcoming shots by touching and holding the focus area until you see AE/AF Lock; tap on the screen to unlock settings.

You can accurately set and lock the exposure settings for upcoming shots on iPhone 11 and iPhone 12. Tap ⚫, tap ⊕, then adjust the exposure by moving the slider. The exposure locks until you close and reopen the Camera. To stop the exposure control from resetting whenever you open Camera,

Go to **Settings** ⚙ > **Camera** > **Preserve Settings**, then turn on **Exposure Adjustment**.

Taking low-light photos with Night mode

Night mode captures additional details and brightens your shots in low-light situations (available on iPhone 12 models and iPhone 11 models). The iPhone camera automatically determines the length of the exposure, but you can explore the manual controls.

Night mode is available on both the front camera and the rear cameras (Ultra Wide (0.5x) camera and Wide (1x) camera) on the iPhone 12 models. While on the iPhone 11 models, Night mode is only available on the rear camera (Wide (1x) camera).

Night mode automatically turns on during low-light situations

- On Photo mode, Night mode automatically turns on during low-light situations. The ⊚ button at the top left of the screen turns yellow, and a number appears next to the ⊚ button to specify how many seconds the camera will take to shoot.

- To explore Night mode manual controls, tap ⊚, then to choose between the Auto and Max timers, use the slider below the frame. The time is automatically determined if you choose Auto; Max uses the longest time. Your customized setting is preserved for your next Night mode shot.

- Tap the Shutter button, then hold the iPhone still to take your shot.

If your iPhone detects movement during capture, crosshairs appear. Align the crosshairs to reduce motion and improve the shot. Tap the Stop button below the slider to stop taking a Night mode shot in mid-capture.

Taking a Live Photo

A Live Photo captures what happens 1.5 seconds before and after taking your photo, including the audio. What you capture is more than a great photo; it is a moment captured with movement and sound.

- In Photo mode, tap ⊚ to turn Live Photos on or off.
- Then tap the Shutter button to take the shot.

You can also add effects to Live Photos, e.g., Loop and Bounce. *See* Edit Live Photos under Photos (page 88).

Taking a photo with a filter

1. In Photo or Portrait mode, do one of the following:

- On iPhone XS, iPhone XR, and later (to iPhone 12), tap ⬆, then tap ⚙.
- On iPhone X and earlier, tap ⚙ at the top of your screen.

2. Below the viewer, preview the filters by swiping them left or right; tap a filter to choose it.

In Photos, you can remove or change a photo's filter. *See* Reverting an edited photo (page 87).

Taking Burst shots

Burst mode takes multiple high-speed shots so that you have a range of pictures to choose from. Burst shots is available in both the front and rear cameras.

1. On iPhone XS, iPhone XR, and later (to iPhone 12), swipe the Shutter to take rapid-fire shots. While on iPhone X and earlier, touch and hold the Shutter button. The counter displays how many shots you've taken.

2. To stop, lift your finger.

3. To select the photos you want to save, tap the **Burst thumbnail** > **Select**. The suggested photos to keep are indicated with the gray dots below the thumbnails.

4. Tap the circle in the lower-right corner of each photo you wish to save as an individual photo, then tap **Done**. To delete the entire Burst shots, tap the thumbnail > 🗑.

💡 **TIP:** You can quickly take Burst shots by holding the volume up button. Go to **Settings** ⚙ > **Camera** > then turn on **Use Volume Up for Burst** (On iPhone XS, iPhone XR, and later).

Taking a panorama photo

Pano mode enables you to capture landscapes or other shots that won't fit on your camera screen.

*   Swipe the photo screen to choose Pano mode,
*   Pan your iPhone slowly in the direction of the arrow, keeping it on the centerline.
*   Tap the Shutter button again to finish your shot.

💡 **TIP:** To pan in the opposite direction, tap the arrow. Rotate your iPhone to landscape orientation to pan vertically. You can also reverse the direction of a vertical pan.

Navigating the camera's video screen

You can use the camera 📷 to record videos on your iPhone and modes to take time-lapse and slow-motion videos.

The iPhone 12 models feature Dolby Vision, which is a professional HDR video format that provides a more accurate life-like color along with brighter highlights and darker shadows.

Recording a video with your iPhone camera

1. Swipe and choose Video mode, then tap the Record button or press either volume button to start recording. You can do the following while recording:

   - You can snap a still photo by pressing the white Shutter button.
   - Pinch your camera screen to zoom in or out.
   - On iPhone models with Dual (e.g., iPhone 12 and iPhone 12 mini) and Triple (e.g., iPhone 12 Pro and iPhone 12 Pro Max) camera systems, touch and hold 1x, then drag the slider to the left.

2. To stop recording, tap the Record button or press either volume button.

Video records at 30 fps (frames per second) by default. Depending on your iPhone model, you can pick other video resolution and frame rates settings in **Settings** ⚙ > **Camera** > **Record Video**.

IPhone XS, iPhone XR, and later (i.e., to iPhone 12) make use of multiple microphones to attain stereo sound. If you want to turn off stereo recording, go to **Settings** ⚙ > **Camera** > then turn off **Record Stereo Sound.**

iPhone 12 models record video in Dolby Vision HDR (up to 4k at 30 fps on iPhone 12 models and up to 4k at 60 fps on iPhone 12 Pro models) and shares HDR videos with devices using iOS 13.4, macOS 10.15.4, iPadOS 13.4, or later; other devices with older versions receives an SDR version of the same video. If you want to turn off HDR recording, go to **Settings** ⚙ > **Camera** > **Record Video**, then turn off **HDR Video**.

Recording a QuickTake video

You can record a QuickTake video on iPhone XS, iPhone XR, and later (to iPhone 12). QuickTake video allows you to record a video without switching out of Photo mode. While recording a QuickTake video, you can still take still photos by moving the Record button into the lock position.

- While in Photo mode, to start recording a QuickTake video, press and hold the Shutter button.
- For hands-free recording, slide the Shutter button to the right and let go over the lock. Both the Shutter and Record buttons appear below the frame. To take still photos while recording, tap the Shutter button.
- To stop recording, tap the Record button.

Quickly start recording a QuickTake video in Photo mode

💡 **TIP:** You can quickly start recording a QuickTake video in Photo mode by pressing and holding either volume button.

In the Photos app, tap the thumbnail to view the QuickTake video.

Using quick toggles to change video resolution and frame rate

In Video mode, you can use quick toggles at the top of the camera screen to change the frame rates and video resolution available on your iPhone.

On iPhone XS, iPhone XR, and later (to iPhone 12), tap the quick toggles at the top of the camera screen to switch between 24, 30, or 60 frames per second (fps) and between HD or 4K recording in Video mode.

To enable quick toggles on iPhone X and earlier, go to **Settings** ⚙ > **Camera** > **Record Video**, then toggle on Video Format Control.

Recording a slow-motion video

Your video records as usual when you record a video in Slo-mo mode. You only see the slow-motion effect when you play it back. iPhone also allows you to edit your videos; you can choose when the slow-motion effect starts and stops.

- Swipe and choose **Slo-mo mode**. You can tap the

  icon 🔄 to record in Slo-mo mode with the front camera on iPhone 12 and iPhone 11 models.

- Press either volume buttons or tap the Record button to start recording. You can take still photos while recording by tapping the Shutter button.
- To stop recording, tap the Record button or press either volume button.

To edit the video, and set the parts of the video you want to play in slow motion and the rest at normal speed, tap the video thumbnail, then tap **Edit**. To define the section you want to playback in slow motion, slide the vertical bars below the frame viewer.

You can change the slow-motion frame rate and resolution depending on your model. To change it, go to **Settings** ⚙ > **Camera** > **Record Slo-mo**.

💡 **TIP:** You can quickly adjust the frame rate and video resolution while recording by using quick toggles. See the previous heading.

Capturing a time-lapse video

The time-lapse video involves capturing lots of photos of a scene (e.g., the setting sun or traffic flowing) over a time period. iPhone then assembles these images to create a seamless sped-up video footage.

- Swipe and choose **Time-lapse mode**.

- Then set up and position your iPhone camera at the scene in motion you want to capture.
- To start recording, tap the Record button; then tap it again when you are through to stop recording.

TIP: On iPhone 12 models, use a tripod to capture a professional-looking time-lapse video when recording in low-light situations.

Adjusting Auto FPS settings

IPhone XS, iPhone XR, and later (to iPhone 12) can automatically reduce the frame rate to 24 fps to improve the video quality in low-light situations.

1. Go to **Settings** ⚙ > **Camera** > **Record Video**, and do either of the following:
   - On iPhone 12 models, tap **Auto FPS**, then apply it to only 30-fps video or to both 30- and 60-fps video.
   - While on iPhone XS, iPhone XR, iPhone SE (2nd generation), and iPhone 11 models, turn on **Auto Low Light FPS**.

Taking Portrait mode photos on iPhone

On the iPhone 7 Plus, iPhone 8 Plus, iPhone X, and later (iPhone 12) Camera 📷, you can apply a depth-of-field effect that will keep your subject perfectly sharp while creating a beautifully blurred background. You can also apply and adjust different lighting effects to your Portrait mode photos, and on iPhone X and later (iPhone 12), you can even apply the Portrait mode effect on selfies.

Taking a photo in Portrait mode

You can apply studio-quality lighting effects to your Portrait mode photos on iPhone 8 Plus, iPhone X, later (iPhone 12).

You can apply studio-quality lighting effects to your Portrait mode

1. Swipe and choose **Portrait mode**.
2. To frame your subject in the yellow portrait box, follow the tips onscreen.
3. Then choose a lighting effect by dragging ⬡:

- **Natural Light**: The face will be perfectly sharp, while the background will be beautifully blurred.
- **Studio Light**: The face will be brightly lit, and the photo has an overall clean look.
- **Contour Light**: The face will have dramatic shadows with highlights and lowlights.
- **Stage Light**: The face is spotlit, while the background will be deep black.
- **Stage Light Mono**: This effect is similar to Stage Light, but the photo is in classic black and white.
- **High-Key Light Mono**: On iPhone XS, iPhone XR, and later (to iPhone 12), with this, you can create a grayscale subject on a white background.

4. To take the shot, tap the **Shutter button**.

Adjusting Depth Control in Portrait mode

On iPhone XS, iPhone XR, and later (to iPhone 12), you can adjust the level of background blur in your Portrait mode by using the Depth Control slider.

Drag the slider to the right or left
To adjust the effect

- Swipe and choose **Portrait mode**, then frame the subject.

- In the top-right corner of the screen, tap ƒ. The Depth Control slider appears below the frame.

- To adjust the effect, drag the slider to the right or left.

- To take the shot, tap the **Shutter button**.

Adjusting Portrait Lighting effects in Portrait mode

On iPhone XS, iPhone XR, and later (to iPhone 12), you can virtually adjust the intensity and position of each Portrait Lighting effect to add features like sharpen eyes or brighten and smooth facial.

- Swipe to choose **Portrait mode**, then choose a lighting effect by dragging ⬡.

- At the top of the screen, tap ◉. The Portrait Lighting slider appears below the frame.

- To adjust the effect, drag the slider to the right or left.

- To take the shot, tap the **Shutter button**.

Using Camera settings on iPhone

Learn how to use your iPhone Camera 📷 settings.

Aligning your shots

This feature helps you to straighten your shots by displaying a grid on the camera screen. Go to **Settings** ⚙ > **Camera**, and turn on **Grid**.

Preserving your camera settings

This feature preserves the camera mode, lighting, filter, depth, and Live Photo settings you last used so that when next you open the Camera, it won't reset.

1. Go to **Settings** ⚙ > **Camera** > **Preserve Settings**.
2. Then turn on any of the following:

   - **Camera Mode**: Turn this on to preserve the last camera mode you used, e.g., Pano or Video.

- **Creative Controls**: Turn this on to preserve the last settings you used for the lighting option (available on iPhone 8 Plus, iPhone X, and later), depth control (On iPhone XS, iPhone XR, and later), or filter
- **Exposure Adjustment**: Turn this on to preserve the last exposure control settings (on iPhone 11 and iPhone 12).
- **Live Photo**: Turn this on to preserve the last Live Photo setting.

Mirroring the front camera

This feature enables you to take a mirrored selfie that captures the shot exactly as you see it in the camera frame. To turn this feature on, go to **Settings** ⚙ > **Camera**, then turn on **Mirror Front Camera** (available on iPhone XS, iPhone XR, and later).

Pressing volume up for Burst

To quickly take Burst shots by pressing and holding the volume up button. Go to **Settings** ⚙ > **Camera** > then turn on **Use Volume Up for Burst** (On iPhone XS, iPhone XR, and later).

Adjusting the shutter sound volume

You can adjust the shutter sound by pressing the volume buttons on the side of your iPhone. Or, while in the Camera app, you can open Control Center by swiping down from the top-right of the camera screen, then drag ◀)).

You can mute the sound by using the Ring/Silent switch on the side of your iPhone. (muting is disabled in some countries or regions)

N.B. When Live Photo ◉ is turned on, the camera shutter doesn't make a sound.

Turning prioritize faster shooting on and off

On iPhone XS, iPhone XR, and later (to iPhone 12), this feature helps to modify how images are processed so that you can shoot even faster when you rapidly tap the Shutter button.

Prioritize Faster Shooting is on by default. To turn it off, go to **Settings** ◉ > **Camera**, then turn off **Prioritize Faster Shooting.**

Turning lens correction on and off

This feature helps to adjust photos taken with the Ultra-Wide camera or front camera for more natural-looking results (available only on iPhone 12 models).

Lens Correction is on by default. To turn it off, go to **Settings** ⊚ > **Camera**, then turn off **Lens Correction**.

Turning scene detection on and off

This feature helps identify objects in various scenes and automatically improve them (available only on iPhone 12 models).

Scene Detection is on by default. To turn it off, go to **Settings** ⊚ > **Camera**, then turn off **Scene Detection**.

Turning view content outside the frame on and off

This feature helps display the camera preview content outside the frame to show you what can be captured by using another camera lens with a wider field of view (available only on iPhone 12 and iPhone 11 models).

View Outside the Frame is on by default. To turn it off, go to **Settings** ⊚ > **Camera**, then turn off **View Outside the Frame**.

Adjusting HDR camera settings on iPhone

iPhone Camera features HDR (high dynamic range), which helps you get great shots in high-contrast situations. It allows the iPhone to take several photos in quick succession at different exposures and blends them to bring more shadow and highlight details to your photos.

iPhone takes photos in HDR by default (for both the front and rear camera) when it is most effective. iPhone 12 models record video in Dolby Vision HDR to capture true-to-life color and contrast.

Turning off automatic HDR

To turn off automatic HDR and manually control HDR instead, do one of the following:

- On iPhone XS, iPhone XR, and later (to iPhone 12), go to **Settings** ⚙ > **Camera**, then turn off **Smart HDR**. Then from the camera screen, you can manually control HDR; tap it to turn it off or on.

- On iPhone 8, iPhone 8 Plus, and iPhone X, go to **Settings** ⚙ > **Camera**, then turn off **Auto HDR**. On the camera screen, to turn HDR back on, tap **HDR**, then tap **On**.

- On iPhone 7, iPhone 7 Plus, and earlier, at the top of the camera screen, tap **HDR**, then tap **Off**.

Keeping the non-HDR version of a photo

The HDR version of a photo is saved in Photos by default. You can also save the non-HDR version (on iPhone X and earlier models). Go to **Settings** ⚙ > **Camera**, and turn on **Keep Normal Photo**.

Turning HDR video on and off

On iPhone 12 models, iPhone records video in Dolby Vision HDR (up to 4k at 30 fps on iPhone 12 models and up to 4k at 60 fps on iPhone 12 Pro models) for true-to-life color and contrast.

HDR video recording is on by default. To turn it off, go to **Settings** ⚙ > **Camera** > **Record Video**, then turn off **HDR Video**.

Viewing, sharing, and printing photos on iPhone

All photos and videos taken with the iPhone Camera 📷 are saved up in Photos.

Viewing your photos

- In camera, to view a photo, tap the thumbnail image in the lower-left corner.
- To see other photos you have taken recently, swipe right or left.
- To show or hide the controls, tap the screen.

- To see all your saved photos and videos in Photos, tap **All Photos**.

Sharing and printing your photos

- Tap ⬆ while viewing a photo.
- Choose an option, e.g., AirDrop, Mail, or Messages, to share your photo.
- Swipe up, then select Print from the list of actions.

Scanning a QR code with the iPhone camera

You can use the iPhone camera to scan Quick Response (QR) codes for the links to tickets, coupons, apps, websites, etc.

Using the camera to read a QR code

- Open the **Camera**, then position your iPhone camera so that the code appears on the screen.
- Tap the notification that pops up on your screen to go to the relevant app or website.

4: Photos

The Photos app 🌸 is a simple tool that enables you to view all the visual media (photos and video) stored on your iPhone. The app allows you to view, edit, and manage all your photos and videos stored up in iPhone.

Viewing photos and videos on iPhone

You can use the Photos app to find and view photos and videos quickly. The app organizes your photos and videos by years, months, days, and all photos. To create albums you can share with family and friends, and to find photos organized by different categories, tap the **For You**, **Albums**, and **Search tabs**.

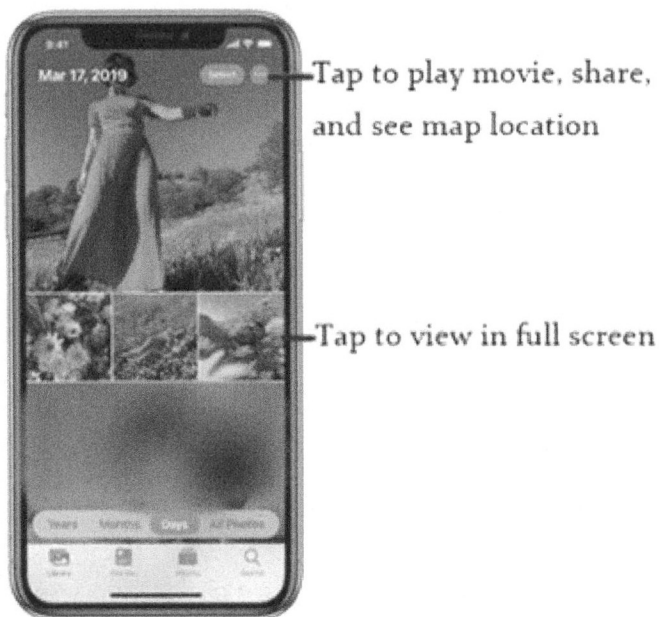

Tap to play movie, share, and see map location

Tap to view in full screen

- **Library**: Tap this tab to browse your photos and videos organized by days, months, and years.
- **For You**: Tap this tab to view a personalized feed that shows your memories, shared albums, featured photos, etc.
- **Albums**: Tap this tab to view albums you created or shared, and also to view your photos organized by album categories (e.g., People & Places and Media Types).
- **Search**: Tap this tab to search for photos by caption, date, location, or objects they contain.

Browsing your photos

In the Library tab, the photos and videos on your iPhone are organized by Years, Months, Days, and All Photos. This enables you to rediscover your finest shots in Years, relive important moments in Months, focus on exceptional photos in Days, and view every photo in All Photos.

- To surf through your photos, tap the **Library tab**.
- Then tap either **Years**, **Months**, **Days**, or **All Photos**.

Viewing individual photos

You can locate and view individual photos in the Library tab. To view individual photos, tap All Photos, or browse by Days, Months, or years, then click on a photo thumbnail to view it in full screen. You can also view individual photos in the Albums tab or tap the Search tab to look for photos by specific search terms, e.g., "waterfall" or "selfie."

Swipe to scroll through your photos

You can do the following while viewing a photo in full screen:

- **Zoom in or out**: Pinch out or double-tap to zoom in. While zoomed in, you can drag to see other parts of the photo. Then to zoom out, pinch close, or double-tap.

- **Share**: Tap ⬆, then choose a sharing option. *See Sharing Photos and videos on iPhone (page 99).*

- **Add to favorites**: To add the photo to your Favorites album in the Albums tab, tap ♡.

To continue browsing or to return to the search results, tap ‹ or drag the photo down.

Adding captions and view photo and video details

Use Captions to add context to your videos and photos, and you can look for videos and photos by captions when you use Search. To add a caption, pick a video or photo and swipe up to view a caption, or you can add a caption in the text field below the image.

You see the following details when you swipe up on a photo or video:

- Effects you can add to a Live Photo; *see* Editing a Live Photo (page 89).
- People identified in your photo; *see* Finding people in Photos (page 104).
- Where the photo was taken; *see* Browsing photos by location on iPhone (page 106).
- A link to view other photos that were taken nearby.

Playing a Live Photo

A Live Photo ◉ is simply a moving image that captures what happens 1.5 seconds before and after taking your photo.

- Open a Live Photo, then touch and hold the photo to play it. *See* Taking a Live Photo (page 55).

Viewing photos in a Burst shot

Burst mode takes multiple high-speed shots so that you have a range of pictures to choose from. In Photos, your Burst shots are saved together in a single photo thumbnail.

- Open a Burst photo, tap **Select**, then go through all the photos.
- To save specific photos, tap and select each photo, then tap **Done**.
- To keep both the Burst and the photos you selected, tap **Keep Everything**, or tap **Keep Only Favorites** to keep only the ones you selected.

See Taking Burst shots (page 56) for more information.

Playing a video

Videos auto-play while you scroll when browsing your photo library in the Library tab. Tap a video to play it in full screen without sound, then do any of the following:

- Below the video, tap the player controls to pause, play, unmute, mute; tap the screen to hide the player controls.

- To toggle between fit-to-screen and full screen, double-tap the screen.

The plaayer controls

Playing and customizing a slideshow

You can create a slideshow with your iPhone.

- Tap the **Library tab**.
- View photos by **Years**, **Months**, **Days**, or **All Photos**, then tap **Select**.
- Click on the photos you want to include in the slideshow, then tap ⬆.
- Tap **Slideshow** from the list of options.

- Tap the screen, then in the bottom right, tap Options to change the slideshow theme, music, etc.

Deleting and hiding photos and videos on iPhone

You can delete photos and videos, hide them in the Hidden album, or recover photos you recently deleted in the 🌼 Photos app

Deleting or hiding a photo or video

In Photos, click on a photo or video and do one of the following:

- **Delete**: To delete a photo or video from your iPhone, tap 🗑. Your deleted photos and videos are kept for 30 days in the Recently Deleted album. You can recover or permanently delete them from the album.

- **Hide**: Tap ⬆, then in the list of options, tap **Hide**. Your hidden photos and videos are moved to the Hidden album, where you can view them. To stop Hidden album from appearing in Albums, go to **Settings** ⚙ > **Photos**, then turn off **Hidden album**.

Recovering or permanently deleting a deleted photo

- Go to the **Albums tab**, then under Utilities, tap **Recently Deleted**.

- Tap **Select**, then choose the photos and videos you wish to recover or permanently delete.

- Tap either **Recover** or **Delete** at the bottom of the screen.

Editing photos and videos on iPhone

In the Photos app , there are cool photo and video editing tools at your disposal.

Adjusting light and color

- In Photos, to view a photo or video in full screen, tap the thumbnail.

Drag the slider

- Tap **Edit**, then to view the editing buttons for each effect such as Highlight, Exposure, and Brilliance, swipe left under the photo.

- Tap an editing button, then adjust the effect by dragging the slider. The outline around the button displays the level of adjustment you make for each effect so that at a glance, you can see which effects have been increased or decreased.

- To preview the effect, tap the effect button to see the photo or video before and after applying it (or tap the photo or video to toggle between the edited and original version).

- Tap **Done** to save your edits, or tap **Cancel** if you don't like your changes, then tap **Discard Changes**.

💡 **TIP:** To automatically edit your photos or videos with effects, tap ⊗.

Rotating, cropping, or flipping a photo or video

1. In Photos, to view a photo or video in full screen, tap the thumbnail.

2. Tap **Edit**, then tap ⊡ and do any of the following:

   - **Rotate**: To rotate the photo 90 degrees, tap ⊡

   - **Crop manually**: Drag the rectangle corners to enclose the photo or video and remove unwanted outer areas.

   - **Crop to a standard preset ratio**: Tap ⊟, then choose a ratio, e.g., Square, 2:3, 8:10, etc.

   - **Flip**: To flip the image horizontally, tap ⊼.

3. Tap **Done** to save your edits, or tap **Cancel** if you don't like your changes, then tap **Discard Changes**.

Adjusting and straightening perspective

- In Photos, to view a photo or video in full screen, tap the thumbnail.

- Tap **Edit**, then tap ⊡.

- To adjust vertical or horizontal perspective and to straighten, select an effect button.
- Then adjust the effect by dragging the slider. The yellow outline around the button displays the level of adjustment you make for each effect so that at a glance, you can see which effects have been increased or decreased. To toggle between the edited effect and the original, tap the button.

Drag the slider
to tilt or straighten

- Tap **Done** to save your edits.

Applying filter effects

- In Photos, to view a photo or video in full screen, tap the thumbnail.
- Tap **Edit**, tap 🎨 to apply filter effects such as Dramatic, Vivid, Silverstone.
- Tap the photo to compare the edited photo to the original.
- Tap **Done** to save your edits, or tap **Cancel** if you don't like your changes, then tap **Discard Changes**.

Reverting an edited photo

You can revert an edited photo to the original image.

- Open the edited photo, tap **Edit** > **Revert**.
- Then tap **Revert to Original**.

Marking up a photo

- In Photos, to view a photo in full screen, tap the thumbnail.
- Tap **Edit**, tap 🔘, then tap **Markup** Ⓐ.
- Use different drawing tools and colors to annotate the photo. To add texts and shapes, tap ＋.

Trimming a video

- In Photos, to view a video in full screen, tap the thumbnail.
- Tap **Edit**, then trim the video by dragging either end of the frame viewer.

Drag either end
of the frame viewer

- Tap **Done**, then tap **Save Video as New Clip** to save both versions of the video, or tap **Save Video** to save only the trimmed video.

Editing Live Photos on iPhone

In the Photos app 🌼, you can edit a Live Photos by changing the Key Photo, adding cool effects like loop and bounce, etc.

Editing a Live Photo

Tap to make a still photo

1. Open the Live Photo and tap **Edit**.

2. Then tap ◎ and do any of the following:

    • **Setting a key photo**: Move the white frame on the frame drawer, then tap **Make Key Photo** > **Done**.

    • **Trimming a Live Photo**: To choose the frames the Live Photo plays, drag either end of the frame drawer.

    • **Making a still photo**: At the top of the screen, tap the Live button to turn off the Live

Feature. The Live Photo becomes a still of its Key photo.

- **Muting a Live Photo**: At the top of the screen, tap ◀))). Tap it again to unmute.

See Taking a Live Photo (page 55) for more information.

Adding effects to a Live Photo

In Photos, you can add fun effects to Live Photos.

1. Open the Live Photo.
2. To see the effects swipe up, then do one of the following:
    - **Loop**: This effect repeats the action in a continuous looping video.
    - **Bounce**: This effect rewinds the action backward and forward.
    - **Long Exposure**: This effect stimulates a DSLR-like long exposure effect by blurring motion.

Editing Portrait mode photos on iPhone

In the Photos app 🌸, you can change and adjust the lighting effects and depth control of the photos you take in Portrait mode.

Editing Portrait Lighting effects in Portrait mode photos

You can apply, change, or remove studio-quality lighting effects to your Portrait mode photos on iPhone 8 Plus, iPhone X, to later (iPhone 12).

1. In Photos, tap the thumbnail of any photo taken in Portrait mode to view it in full screen.

Drag the slider to fine-tune the intensity of the lighting effect

2. Tap **Edit**, then choose a lighting effect by dragging ⬡ below the photo:

  • **Natural Light**: The face will be perfectly sharp, while the background will be beautifully blurred.

- **Studio Light**: The face will be brightly lit, and the photo has an overall clean look.
- **Contour Light**: The face will have dramatic shadows with highlights and lowlights.
- **Stage Light**: The face is spotlit, while the background will be deep black.
- **Stage Light Mono**: This effect is similar to Stage Light, but the photo is in classic black and white.
- **High-Key Light Mono**: On iPhone XS, iPhone XR, and later (to iPhone 12), with this, you can create a grayscale subject on a white background.

  N.B. Only the front camera supports Natural Light, Studio Light, and Contour Light on iPhone XR.

3. To fine-tune the intensity of the lighting effect, drag the slider left or right.
4. To save your changes, tap **Done**.

Tap **Edit** > **Revert** to undo the Portrait Lighting effect after you save.

N.B. Tap **Portrait** at the top of the screen to remove the Portrait effect from a photo. *See* Taking Portrait mode photos on iPhone (page 63) for more information.

Adjusting Depth Control in Portrait mode photos

On iPhone XS, iPhone XR, and later (to iPhone 12), use the Depth Control slider to fine-tune the level of background blur in your Portrait mode.

Use the Depth Control slider to adjust the level of background blur

- In Photos, tap the thumbnail of any photo taken in Portrait mode to view it in full screen.

- Tap **Edit**, then at the top corner of the screen, tap *f*. The Depth Control slider appears below the frame.

- To adjust the effect, drag the slider to the right or left.
- To save your changes, tap **Done**.

See Taking Portrait mode photos on iPhone (page 63) for more information.

Organizing photos in albums on iPhone

You can use the Photos app 🌸 to organize your photos with albums. Tap the **Albums tab** to view albums you created, albums created automatically, and Shared Albums you created or joined.

Tap to create
an album

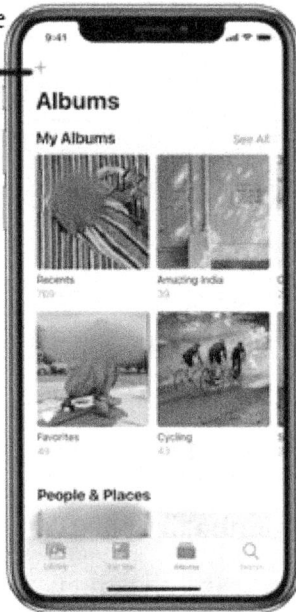

New Folder

Creating an album

- In Photos, tap the Album tab, then tap +.
- Choose either create a New Album or New Shared   ?
  Album.
- Write a name for your album, then tap **Save**.
- Select the photos you want to add to the album, then
  tap **Done**.

Adding photos and videos to existing albums

- At the bottom of the screen, tap the **Library tab** >
  Select.
- Tap the thumbnails of the photo and video you want
  to add, then tap ⬆.
- Swipe up, then from the list of actions, tap **Add to**
  Album.

- Then choose the album you want to add it to.

Removing photos and videos from existing albums

- While in an album, tap a photo or video to view it in full screen.

- Then tap 🗑.

Renaming, rearranging, and deleting existing albums

1. Tap the **Album tab** > **See All** > **Edit**.

2. Then do any of the following:

   - Renaming: Tap the name of the album, then type a new name.

- Rearranging: Touch and hold the album's thumbnail, then drag it to a different location.
- Deleting: Tap ⊖.

3. Tap **Done**.

You cannot delete the albums that Photos created for you, e.g., Recents, People, and Places.

Sorting photos in albums

In an album, you can sort your photos and videos by newest or oldest, or oldest to newest.

- Tap the **Albums tab**, then select an album.

- Tap ⋯ > **Sort**.

Searching for photos on iPhone

In the Photos app 🌼, when you click on the Search tab, iPhone helps you locate whatever you search for or rediscover a moment you forgot about by giving you suggestions for people, places, and photo categories. You can type a keyword into the search bar, e.g., a person's name, date, or location, to help you find a specific photo.

🎙️**Ask Siri.** Say something like: "Show me photos from June 2016." *See* Learn how to talk to Siri (page 219).

Tap the search field

Tap the **Search tab**, then at the top of the screen, tap the search field to search by the following:

- Date (month or year)
- Place (city or state)
- Business names (shopping malls, for example)
- Category (sunset or beach, for example)
- Events (concerts or sports games, for example)
- Person (Eve or Matt, for example)
- Caption (*see* Adding captions and view photo and video details (page 79))

Sharing photos and videos on iPhone

In the Photos app ⬡, you can share photos and videos in Mail or Messages, or other apps you install. The photos app even gives you suggestions of your best shots from an event and also recommends people to share them with.

Sharing photos and videos

- **Sharing a single photo or video**: Tap the thumbnail to open the photo or video, tap ⬆, then choose a share option.

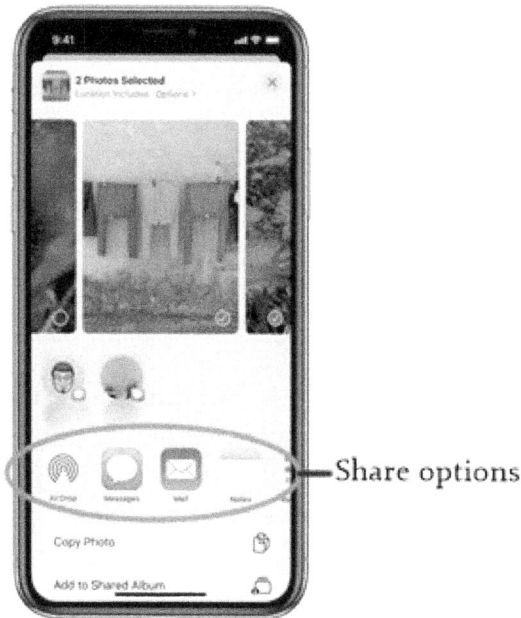

Share options

- **Sharing multiple photos or video**: While viewing photos in the All Photos tab in your Library or an album, tap **Select**, and tap the photos you want to share. Tap ⬆, and choose how you want to share it.
- **Sharing photos or videos from a Day or Month**: While in your Library, tap the **Days or Months tab**, tap ●, then to share all the photos from that day or month, tap **Share Photos**.

Using Sharing Suggestions to share photos

Sharing suggestions assist you by recommending a set of photos from an event that you may want to share based on the people in the photos and who may want to share the photos with you.

After you've shared your photos from the event, the recipients are prompted to share theirs with you. Sharing Suggestions requires iCloud Photos enabled to share.

- Tap the **For You tab**, then in Sharing Suggestions, tap a photo collection.
- Tap **Select** to add or remove photos > **Next**.
- Tap **Share in Messages**.

Saving or sharing a photo or video you received

- **From a text message**: In the conversation, tap the photo or video, tap ⬆, then choose a saving or sharing option.

- **From email**: If necessary, tap to download the item, then tap ⬆. Or touch and hold the item, then select a saving or sharing option.

Viewing Memories in Photos on iPhone

The Photos app 🌸 automatically creates collections of photos and videos called Memories by scanning your Library. Memories comprise of photos and a movie, which is edited automatically and set to music for you. You are allowed to edit a Memory movie and share it with friends. You are also allowed to make a Memory from an album you create.

Playing a Memory movie

- Tap the **For You tab**, then select a Memory.
- At the top of the screen, tap the video to play it.
- Tap the movie, then tap ❙❙ to pause it.

- Tap the movie, then slide the frames at the bottom of the screen left or right to go backward or forward in the movie.

Tap ⬤ in the top-right corner of the event and tap Play Memory Movie to see a Memory movie from an event in Days or Months.

Editing a Memory movie

1. Tap the screen to show customization options while a Memory movie is playing.

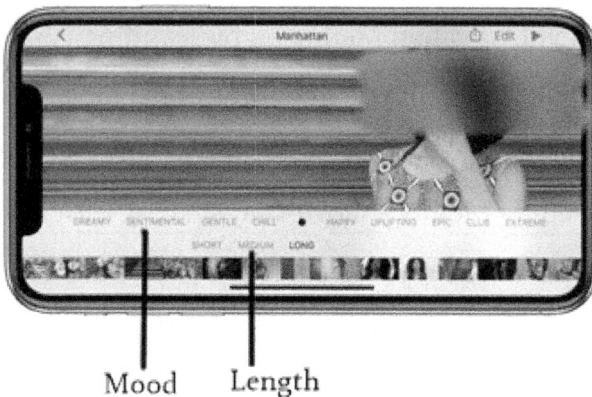

Mood   Length

2. Then do any of the following:

- **Changing the length**: Depending on the number of photos in the Memory, swipe to change to short, medium, or long.

- **Changing the mood**: To change the music and editing style, swipe left or right through the moods.

- **Personalizing the movie**: At the top of the screen, tap Edit to add or delete specific photos, edit the title, change the music, etc.

3. Tap ▶ play.

Creating a Memory

- In the Library tab, view photos by Days or Months, then tap ⬤. Or in the Albums tab, open an album, then tap ⋯ .

- Tap **Play Memory Movie**, then while the movie plays, tap the screen, then tap **Edit**.

- Tap **OK**, then edit the movie by adding or deleting specific photos, editing the title, and changing the music.

- Tap **Done**.

Deleting a Memory

- Tap the **For You tab**, then select a Memory.

- Tap ⬤, and select **Delete Memory** or **Suggest Fewer Memories** Like This.

Sharing a Memory

- Tap the **For You tab**, then play the Memory you wish to share.

- While it is playing, tap the screen, tap ⬆, then choose a sharing option.

Sharing photos from a Memory

Photos identify the people in a Memory. Thus, making it easy to share the photos in that Memory with them.

- Tap the **For You tab**, then tap a Memory with photos you wish to share.

- Tap **Select**, then tap the photos you wish to share.

- Tap ⬆, then choose a sharing option.

Adding a Memory to Favorites

- Tap the **For You tab**, then tap a Memory you wish to mark as favorite.

- Tap ⚫, then tap **Add to Favorite Memories**.

Tap **See All** > **Favorites** to view your favorite Memories.

Finding people in Photos on iPhone

The Photos app 🌼 scans your Library for faces and adds the most frequently detected faces to the People album. You

can, therefore, search for photos by name when you add
names to the faces

Finding photos of a specific person

- Tap the **People album** in the Albums tab, then tap a
  person to view all the photos they are found in.
- In the Search tab, type in the person's name in the
  search field.

Naming a person in your People album

- Tap the **People album** in the Albums tab, then tap
  the face of the person whose name you want to add.
- At the top of the screen, tap **Add Name**, then type in
  the person's name.
- Tap **Next** > **Done**.

If a person's face is identified as two or more people in the
People album, tap **Select**, then tap each instance of the
person > **Merge**.

Adding a person to your people album

- Open a photo of the person you wish to add, then
  swipe up to see more details.
- Under **People**, tap a face > **Add Name**.
- Type in the person's name or select it from the list.

- Tap **Next** > **Done**.

Setting a person's key photo

- Tap the **People album**, then select a person.
- Tap **Select** > **Show Faces**.
- Choose the photo that you want to use as the key photo.
- Tap ⬆ > **Make Key Photo**.

Fixing misidentifications

- Tap the **People album**, then select a person.
- Tap **Select** > **Show Faces**.
- Tap the misidentified face.
- Tap ⬆ > **Not this person**.

Browsing photos by location on iPhone

The Photos app 🔘 enables you to view photos taken in a specific location or look for photos taken nearby by creating collections of your photos and videos in the **Places album**, based on where they were taken. You can also watch a Memory movie of a place or view a collection of all your places on a map.

Browsing photos by location

- Tap the **Places album** in the Albums tab.
- Select either Map or Grid view.

N.B. Only photos and videos that have GPS data (location information) are included.

Checking where a photo was taken

- Open a photo, then to see the photo details, swipe up.
- Tap the address or map link to see more details.

Looking for photos taken nearby

- Tap a photo's thumbnail to view it in full screen.
- Swipe up > **Show Nearby Photos**.

Watching a location-based Memory movie

- Tap the **Places album** in the Albums tab > **Grid**.
- Look for a location with several images, then tap the location heading.
- Tap ▶ play.

Importing photos and videos on iPhone

You can import photos and videos from a digital camera, an SD memory card, or another iPhone, iPad, or iPod touch to

the Photos app 🌸. Use the Lightning to SD Card Camera
Reader or the Lightning to USB Camera Adapter (both sold
separately).

1. Insert the card reader or camera adapter into the
    iPhone's Lightning connector.

2. Then do one of the following:

    - **Insert an SD memory card into the card
      reader**: It fits only one way; don't force the
      card into the reader's slot.

    - **Connect a camera**: Connect the camera to
      the camera adapter with the USB cable that
      came with the camera. Then turn on the
      camera and make sure it is in transfer mode.
      See the material that came with the camera for
      more information.

    - **Connect an iPhone, iPad, or iPod touch**:
      Connect the device to the camera adapter with
      the USB cable that came with the device.

3. Open Photos on your iPhone, then tap **Import**.

4. Tap and select the photos and videos you want to
    import, then choose your import destination.

    - Tap **Import All** to import all items

- To import just some items, tap the items you
  want to import (a checkmark appears for each)
  > **Import** > **Import Selected**.

5. You can keep or delete the photos and videos on the
   camera, card, iPhone, iPad, or iPod touch after they
   have been imported.

6. Disconnect the card reader or camera adapter.

**Printing photos on iPhone with an AirPrint-enabled
printer**

You can directly print your photos from the Photos app 🖼
on your iPhone with any AirPrint-enabled device.

- To print a single photo, tap ⬆ while viewing the
  photo > **Print**.

- To print multiple photos, tap **Select** while viewing
  photos, select the photos you want to print, then tap
  ⬆ > **Print.**

5: Phone

Use the phone app to make telephone calls and access other advanced features.

- From your Home screen, tap **Phone app**

Answer or Decline incoming calls on your iPhone

Learn how to answer, silence, and decline incoming calls on your iPhone. It goes directly to voicemail when you decline a call. You can reply with a text or remind yourself to return the call.

Decline   Answer

Answer a call

Do one of the following to answer a call on your iPhone:

- Tap 📞
- Drag the slider if iPhone is locked
- To answer a call with your EarPods, press the center button

💡 **TIP:** To have your iPhone announce all incoming calls or only the calls you receive while you're using Bluetooth in your car or headphones, go to **Settings** ⚙️ > **Phone** > **Announce calls.**

Silence a call

To silence a call on your iPhone, press either the volume button or side button or the Sleep/Wake button (depending on your model).

N.B. You can still answer a silenced call until it goes to voicemail.

Decline a call

Do one of the following to decline a call and send it directly to voicemail:

- Quickly double press the side button or Sleep/Wake button (depending on your model)
- Tap
- Or swipe up on the call banner to decline an incoming call

You can also swipe down on the banner to access these options.

Do one of the following:

- Tap **Remind me** and set when you want a reminder to return the call.

- Tap **Message**, then select either Default reply or Custom.

To make your default replies, Go to **Settings** ⚙ > **Phone** > **Respond with text**, and tap on any default message to replace it with your own.

N.B. Declined calls are disconnected without being directed to voicemail in some countries or regions.

Making a call on iPhone

Make and answer phone calls with your iPhone.

- To make a call in the **Phone app** , you can either dial the number on the keypad or tap a favorite, recents, or choose a number in your contact list.

Dial a number

You can dial the number you wish to call with either **Ask Siri** or the typical way.

Ask Siri: Say "dial" or "call," followed by a number. You should speak each digit separately (i.e., "three, one, two, six, six, four..."). *See* Learn how to talk to Siri (page 219).

Or the common way:

1. From the Phone app's screen, tap **Keypad**.

Tap to make the call on another line

2. The following options are available:

- **Use a different line:** On models with Dual sim, tap the line at the top, then choose a different line to make the call.

- **Enter the number using the keypad**: Dial the number with your keypad. If you make a mistake, tap ⊗ to erase.

- **Redial the last number:** Tap 📞 to view your last dialed number, then tap 📞 to call the number.

- **Paste a number you have copied:** Press and hold the phone number field above the keypad, then tap Paste.
- **Enter a 2-second (soft) pause:** Press and hold the star (*) key until it changes to a comma.
- **Enter a hard pause (to pause dialing until you tap the Dial button):** Press and hold the pound (#) key until it changes to a semicolon.
- **Enter a "+" for international calls:** Press and hold the "0" key until it changes to "+."

3. Tap ⚪ to start the call, and to end the call, tap ⚫.

Making a call from Favorites

1. From the Phone app's screen, tap **Favorites**, then choose one of your Favorites to make a call. On models with Dual sim, iPhone chooses the line for the call in this order:
   - The preferred line for this contact (i.e., if set).
   - The last line used for the last call to or from this contact.
   - The default voice line.

2. Do any of the following to manage your Favorites list:

- **Add a favorite**: From **Favorites**, tap ✛, then choose a contact to add to a Favorite list.
- **Delete or Rearrange favorites**: Tap **Edit**.

Making a call from Contacts

You can make calls from your contacts list with either 🎙 **Ask Siri** or the typical way.

🎙 **Ask Siri:** Say something like "Call Eve's mobile." See Learn how to talk to Siri (page 219) for more details.

Or the common way:

1. From the Phone app's screen, tap **Contacts**.
2. Tap the contact, then tap the number you wish to call.

N.B. On models with dual sim, the default line is used to make the call unless you set a preferred line for this contact.

Making a call from Recents

You can make calls from Recents with either 🎙 **Ask Siri** or the common way.

🎙 **Ask Siri:** Say something like "Return my last call" or "Redial that last number." *See* Learn how to talk to Siri (page 219) for more details.

Or the common way:

1. From the Phone app's screen, tap **Recents**.

2. Tap the one you wish to call.

3. Tap ⓘ to get more info or details about a call and the caller.

N.B. A red badge indicates the number of missed calls.

Change your outgoing call settings

1. Go to your phone **Settings** ⚙ > **Phone**.

2. You have the following options:

    • **Turn on Show My Caller ID**: Your phone number is displayed in My Number. Note that for FaceTime calls, your phone number will be displayed even when the caller ID is turned off.

    • **Turn on Dial Assist for international calls**: Turn on this feature to enable your iPhone to automatically add the correct international or local prefix whenever you make a call.

Contact your carrier for information about making international calls (including rates and other charges that may apply).

Actions while on a call on iPhone

While on a phone call, you can switch the audio to speakers or Bluetooth. If you get a second incoming call, you can respond or ignore it.

Adjust the audio during a call

Press the side volume buttons to adjust the audio. Or swipe down on the banner to access the following:

Swipe down on the banner to access these options

- **Mute**: Tap the mute button on your screen.
- **Put the call on hold**: Press and hold the mute button.

- **Talk handsfree**: Tap the audio button on your screen, then choose an audio destination.

Multitask: Use another app while on a call

- To go to the **Home screen**, swipe up from the bottom edge of your screen (for iPhone with face ID) or press the Home button (for iPhone with Home button), then tap and open an app.
- To go back to the call, tap the green indicator at the top of your screen.

Respond to a second incoming call on the same line

Do one of the following if you receive a second call while on a call:

- **Ignore and send the call to voicemail**: Tap **Ignore.**
- **Answer the new call and end the first call**: On GSM network, tap **End + Accept**. While on CDMA network, tap **End,** and when the second call rings back, tap **Accept** or drag the slider if the iPhone is locked.
- **Put the first call on hold and answer the new one**: Tap **Hold + Accept**. Tap **Swap** to switch between calls, or tap **Merge Calls** to talk to both parties at once.

N.B. If you're using CDMA network, you can't switch between calls if the second call is outgoing, but you can merge the calls. If the second call is incoming, you can't merge the calls. Both calls will be terminated if you end the second call or merge the calls.

Start a conference call

You can set up a conference call with up to five people (depending on your carrier) with the GSM network.

1. While on a phone call, tap **Add Call**, make another call > **Merge Calls**.

    Repeat this process to add more people to the conference.

2. During the conference call, you can do the following:

    - **Talk privately with one person**: click ⓘ, then tap **Private** next to the person. To resume the conference, tap **Merge Calls**.

    - **Add an incoming caller on the same line**: Click **Hold Call + Answer** > **Merge Calls**.

    - **Drop one person**: Tap ⓘ next to a person > **End**.

Checking voicemail on your iPhone

Visual Voicemail is available on your iPhone device. Unlike the sequential listening required in the traditional voicemail, Visual Voicemail shows you a list of your messages. And lets you pick which one to play or delete without listening to them.

- From **Phone app** , **tap Voicemail.**

N.B. A badge on the Voicemail icon indicates the number of unheard messages.

Voicemail transcription (Beta; available only in certain countries or regions) allows you to read your voicemails by showing your messages transcribed into text. Voicemail transcription is limited to voicemails in English received on iPhone with iOS 10 or later. Transcription depends on the quality of your recording.

Voicemail, Visual Voicemail, Voicemail transcription are available from select carriers in select countries or regions.

Setting up voicemail

When you first tap Voicemail, you will be asked to create a voicemail password and record your voicemail greeting.

1. From **Phone app** , tap **Voicemail** > **Set Up Now**.

2. **Create a voicemail password** > **Choose a greeting: Default or Custom**; you can record your greeting if you choose Custom.

Play, delete or share your voicemail messages

To play, delete, or share your voicemail messages, you either

🎤 **Ask Siri** or use the standard way.

🎤 **Ask Siri:** Say something like "Play the voicemail from Eve" or "Do I have any new voicemail?" *See* Learn how to talk to Siri (page 219) for more details.

Or the common way:

1. From **Phone app** 📞, tap **Voicemail** > then tap a message.

2. You have the following options:

    • Tap ▶ to play the message. Messages are saved until you delete them or your carrier erases them.

    • Tap ⬆ to share the message.

    • Tap 🗑 to delete the message.

N.B. In some countries or regions, deleted messages might be permanently erased by your carrier. Changing your sim cards may also delete your voice messages.

To recover your deleted messages, tap **Deleted messages**, tap the message > **Undelete**.

Checking messages when Visual Voicemail is unavailable

- **On your iPhone device**: tap on **Voicemail** and follow the prompts.

- **On another phone**: Dial your phone number, press # or * (depending on your carrier) to skip your greeting, then enter your voicemail password.

Changing your voicemail settings

1. **Change your greeting**: From **Phone app** 📞, tap **Voicemail** > **Greeting**

2. **You can change your voicemail password** by going to **Settings** ⚙ and tap **Phone** > **Change Voicemail password** and insert your new password.

   Contact your carrier if you forgot your voicemail password.

3. **To change the alert for new voicemail**: Go to **Settings** ⚙ > **Sound & Haptics** or **Settings** ⚙ > **Sound.**

Selecting ringtones and vibrations on iPhone

You can use the default ringtone in your iPhone and assign distinctive ringtones to certain people in your contacts. You can also turn the ringer off and use only vibrations.

Changing the alert tones and vibrations

1. Go to **Settings** ⚙ > **Sound & Haptics** (on supported models) or **Settings** ⚙ > **Sound**

2. Drag the slider below Ringers and Alerts to set the volume for all sounds.

3. Tap a sound type, such as ringtone or text tone, to set the tones and vibration patterns for sounds.

4. You have the following options:

   - **Choose a tone** (scroll to see all)

     N.B. Ringtones are used for incoming calls, clock timer, and the clock alarm, while text tones are used for text messages, new voicemail, and other alerts.

   - Tap **Vibration**, then select a vibration pattern, or tap **Create New Vibration** to create your vibration pattern.

Assigning a distinctive ringtone to a contact

1. Go to **Contacts** app .

2. Select a contact > **Edit** > **Ringtone**, then choose a ringtone.

Turning the ringer on or off

On the left side of your iPhone, you'd see the Ring/Silent switch. Flip the switch to put the iPhone in ring mode or silent mode.

N.B. Ring/Silent switch only works on incoming calls and alerts. It has no effect when playing music or videos.

Setting up call forwarding and call waiting on iPhone

Call forwarding is a telephony feature that redirects calls intended for your iPhone to another number, such as your office or home number. You can still make outgoing calls when call forwarding is turned on. Forwarding your calls to a landline phone will help you save airtime on most rate plans.

Call waiting is a smart feature that notifies you of a new incoming call while on a call and enables you to place the first call on hold while answering the second call.

You can set up call forwarding and call waiting on your iPhone if you have a cellular service through a GSM network. Contact your carrier for information about enabling these features if you have cellular service through the CDMA network.

1. Go to **Settings** ◉ > **Phone**.

2. Tap any of the following:

    *   **Call Forwarding**: When call forwarding is on, ↪ appears in the status bar (The row of status icons at the top of the screen that provides information about iPhone). Note that you must be in the cellular network range when you set your iPhone to forward calls; otherwise, calls won't be forwarded.

    *   **Call Waiting**: Incoming calls go to voicemail if you are on a call and call waiting is turned off.

Avoiding unwanted calls on iPhone

This feature allows you to avoid unwanted calls by blocking certain people and sending spam and unknown callers directly to voicemail.

Blocking voice calls, FaceTime calls, and messages from certain people

From **Phone app** 📱, do any of the following:

- Tap **Favorites**, **Recent**, or **Voicemail** > Tap ⓘ next to the number or contact you intend to block, scroll down > **Block this Caller**.

- Tap **Contacts**, tap the contact you intend to block, scroll down > **Block this Caller**.

Managing your blocked contacts

- Go to **Settings** ⚙ > **Phone** > **Blocked Contacts**.
- Then tap **Edit**.

Sending spam and unknown callers to voicemail

1. Go to **Settings** ⚙ > **Phone**.
2. Tap any of the following:

   - **Silence Unknown Callers**: You get notifications for calls only from people in your contacts, recent outgoing calls, and Siri suggestion when it is turned on.

   - **Call Blocking & Identification**: To silence calls identified by your carrier as potential spam or fraud, turn on **Silence Junk Callers** (available with certain carriers).

Making calls using Wi-Fi on iPhone

When your iPhone has a low cellular signal, you can use Wi-Fi calling as an alternative for making and receiving calls via a Wi-Fi network.

1. Go to **Settings** > **Phone** > **Cellular**.
2. Choose a line (below Cellular Plans) if your iPhone has a Dual sim.
3. Tap **Wi-Fi Calling** > **Wi-Fi Calling on This iPhone**.
4. Enter or confirm your address from emergency services.

Making emergency calls on iPhone

Quickly call for help in case of emergency. Use Emergency SOS to swiftly and easily call for help and alert your emergency call.

In the Health app, you can create an emergency Medical ID. iPhone can send this medical information to emergency services when you call or text 911 or use Emergency SOS (for U.S. only)

Dialing emergency number when iPhone is locked

1. Tap **Emergency** from the Passcode screen.
2. Then dial the emergency number (e.g. 911 in the U.S.) > .

Using Emergency SOS

1. To use Emergency SOS, press and hold the side button and either volume button

2. When the Emergency SOS slider appears, continue holding the buttons until iPhone plays a warning sound and starts a countdown (drag the Emergency SOS slider to skip the countdown).

3. When the countdown ends, iPhone calls emergency services. iPhone also alerts your emergency contact that you made a call and sends them your current location (if available) when the emergency call ends.

Changing your Emergency SOS settings

1. Go to **Settings** ⚙ > **Emergency SOS**.

   These settings are available:

   - **Turn Auto Call on or off**: When turned on, iPhone automatically calls the emergency service in your region if you start an Emergency SOS.

   - **Turn the countdown sound on or off**

   - **Manage your emergency contacts**: Tap Edit Emergency Contact in Health or Set Up Emergency Contact in Health

6: Contacts

Use the Contacts app to store and manage your contacts. You can also synchronize with your private account (e.g., WhatsApp).

Adding and using contact information on iPhone

With the **Contacts** app , you can add, view, and edit your contacts list from personal, business, and other accounts.

Ask Siri. Say something like:

- "What is my sister's work address?"
- "Steven Carr is my brother."
- "Send a message to my brother."

Creating a contact

- From the **Contacts** app , tap +.

N.B. Siri also recommends new contacts based on your activities in other apps, such as the emails you receive in Mail and the invitations you receive in Calendar. (To turn

this feature off, go to Go to **Settings** ⚙ > **Contacts** > **Siri &**
Search, then tap **Show Siri Suggestions for Contacts** and
turn off.)

Siri also provides you with contact information
suggestions in other apps based on how you use Contacts
(To turn off this feature, go to **Settings** ⚙ > **Contacts** > **Siri**
& Search, then tap **Learn from this app** and turn off.)

Finding a contact

Enter the contact information like the name, phone,
number, address on the search field at the top of your
contacts list.

Sharing a contact

From your contacts list, tap a contact > Share Contact >
Choose a method for sending the contact information.

Quickly reaching a contact

To quickly send a message, make a phone call or FaceTime
call, compose an email, or send money with Apple Pay, tap
an option below the contact's name in the contact's card.

Set your preferred line for making phone calls and sending SMS/MMS messages (Dual SIM only)

To change the default phone number or email address for a contact, touch and hold the phone number or email.

Deleting a contact

- Go to **contact's card** > **Edit**.
- Scroll down, then tap **Delete**.

Editing contacts on iPhone

With the Contacts app , you can add a birthday, change a label, assign a photo to a contact, etc.

- Tap a contact > **Edit**.
- Then these options are available:

- ○ **Assign a photo to a contact**: Tap **Add photo**, then take a photo or add one from the Photos app.
- ○ **Change a label**: Tap a label, then choose one in the list or create your own by tapping Add Custom Label.
- ○ **Add a birthday, social profile, related name, and more**: Tap ⊕ next to the option.
- ○ **Allow calls or texts from a contact to override Do Not Disturb**: Tap **Ringtone** or **Text tone** > **Emergency Bypass**.
- ○ **Add notes**: Tap the Note field to add notes.
- ○ **Add a prefix, phonetic name, pronunciation, and more**: Tap **Add field** and choose an item in the list.
- ○ **Delete contact information**: Tap ⊖ next to a field.
- • When you are through, tap **Done.**

Adding your contact info on iPhone

You can add your information to your contact card with the Contacts app ⬜. iPhone creates your contact card, called My card with your Apple ID.

Completing My Card

- Tap **My Card** at the top of your contacts list > **Edit**. Contacts help you to set up My Card by suggesting addresses and phone numbers.

- If you didn't find My Card at the top of your contacts list, tap ＋, and enter your information. Then, Go to **Settings** ◉ > **Contacts** > **My Info**, and tap your name in the Contacts list.

Editing My Card

- Tap **My Card** at the top of your contacts list > **Edit**.

Creating or editing your Medical ID

- Tap **My Card** at the top of your contacts list > **Edit** > Scroll down, **Create Medical ID,** or **Edit Medical ID.**

Using other contact accounts on iPhone

Contacts app ▣ allows you to include contacts from other accounts.

Using your Google contacts

- Go to **Settings** ◉ > **Contacts** > **Accounts** > **Google**.
- Sign in to your account and turn on Contacts.

Using your iCloud account

- Go to **Settings** 🕸️ > {your name} > **iCloud** and turn on Contacts.

Adding contacts from another account

- Go to **Settings** 🕸️ > **Contacts** > **Accounts** > **Add Account**
- Select an account, sign in to the account and turn on Contacts.

Accessing a Microsoft Exchange Global Address List

- Go to **Settings** 🕸️ > **Contacts** > **Accounts** > **Exchange**.
- Sign in to your Exchange account and turn on Contacts.

Setting up an LDAP or CardDAV account to access business or school directories

- Go to **Settings** 🕸️ > **Contacts** > **Accounts** > **Add Account** > **Other**.
- Then tap **Add LDAP Account** or **Add CardDAV** and enter the account information.

Keeping contacts up to date across devices

You can use iCloud to keep your contact information up to date across all your devices signed in with the same Apple ID.

- Go to **Settings** ⚙ > {your name} > **iCloud** and turn on Contacts.

Importing contacts from a SIM card (GSM)

- Go to **Settings** ⚙ > **Contacts** > **Import SIM Contacts**.

Using Contacts from the Phone app on iPhone

Use the Phone app on the iPhone to call contacts and add recent callers to the Contacts app 📇.

Adding a favorite

This allows you to put VIP contacts in your Favorites list for quick dialing.

- From **Phone app** 📞, select a contact, then scroll down and tap **Add to Favorites**.

Dial and save a number to Contacts

- From **Phone app** 📞, tap **Keypad**, enter a number, and tap **Add Number**.

- Tap **Create New Contact** or **Add to Existing Contact** and select an existing contact.

Adding a recent caller to Contacts

- From the **Phone app** , tap **Recents**, then tap ⓘ next to the number.
- Tap **Create New Contact** or **Add to Existing Contact**, then select an existing contact.

Hiding duplicate contacts on iPhone

In the Contacts app , link contact cards for the same person in different accounts so they won't appear multiple times in your All Contacts list. Duplicate contacts appear when you have contacts from multiple sources. To keep redundant contacts from showing up in your All Contacts list, contacts from various sources with the same name are linked and presented as a single *unified contact*.

Linking contacts

You can manually unify two entries for the same person if they are not unified automatically.

- Tap one of the contacts > **Edit** > **Link Contacts**.
- Choose the other entry to link to > **Link**.

7: Messages

In the Message app 💬 , send text messages as SMS/MMS via cellular service or with iMessage via Wi-Fi or cellular service. iMessage texts can only be sent to people who use iPhone, iPad, iPod touch, or a Mac. Cellular data rates may apply to texts you send or receive with iMessage (it doesn't count against your SMS/MMS allowances in your cellular data plan).

SMS/MMS texts appear in green bubbles, while iMessage text appears in blue bubbles

- From apps, tap **Messages** 💬 .

Signing in to iMesasage

- Go to **Settings** ⚙ > **Messages**.
- Then turn on iMessage.

Signing in to iMessage on your Mac and other Apple devices using the same Apple ID

All the messages that you send and receive on your iPhone also appear on your other Apple devices if you sign in to iMessage with the same Apple ID on all your devices.

1. On your iPhone, iPad, iPod touch, Go to **Settings** ⚙ > **Messages**.

2. On your Mac, open Messages, then do either one of the following:

   - If you're signing in for the first time, **enter your Apple ID and password** > **Sign in**.

   - If you wish to use a different Apple ID after signing in, select **Messages** > **Preferences**, click **iMessage** > **Sign out**.

Sending a message to a business or group on iPhone

You can send photos, videos, and audio messages to groups of people with the Message app 💬. You can also use the business chat to send a message to a business.

Replying to a specific message in a group conversation

Replying to a specific message in a group conversation improves clarity and helps keep the group conversation organized.

- Touch and hold or double-tap a message in a group conversation, then tap ↰.

- Then type in your response > ⬆.

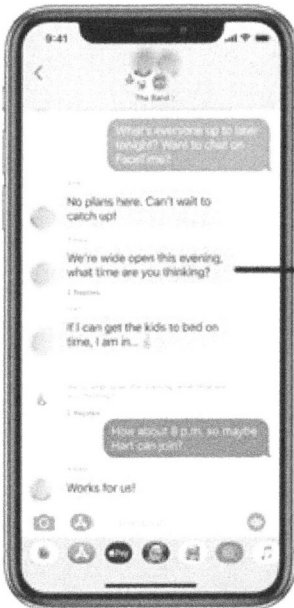

Touch and hold or double-tap to reply a specific message in a group conversation

Mentioning people in a group conversation

You can call other people's attention to a specific message in a group conversation by mentioning them. Depending on their settings, this can notify them of your message even if they have muted the conversation.

- In a group conversation, begin typing a contact's name in the text field.

- Tap the contact's name when it shows on your screen.

N.B. You can also mention a contact in a group conversation by typing "@" followed by the contact's name.

To mention someone in a group conversation, type the contact's name in the text field

You can change your notification settings for when you are mentioned in Messages:

- Go to **Settings** ⑳ > **Messages** > **Notify Me**.

Changing a group name and photo

All the participants are included in the photo used in group conversations, and it changes based on who was recently active. You also have the option to assign a personalized photo to the group conversation.

- Tap the name or number at the top of the group conversation, tap ⓘ at the top right of your screen > **Change Name and Photo**, then choose an option.

Using Business chat

In Messages, you can contact businesses that offer business chat. This allows you to resolve issues, get answers to questions, get advice on what to buy, etc.

- Use Siri, Maps, Safari, or Search to find the business you want to search with.
- You can start a conversation with the business by tapping the chat link in the search result. The chat link can appear in the form of the company logo, 💬 , or a text link.

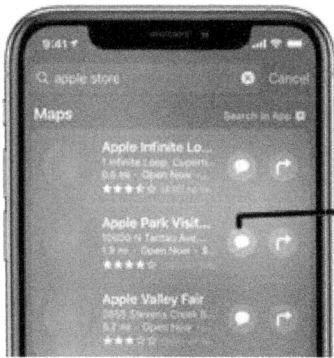

You can start a conversation with the business by tapping the chat link in the search result

N.B. Messages you send in Business Chat appears in dark grey to distinguish them from messages sent using SMS/MMS (in green) and iMessage (in blue)

Sending audios, photos, and video messages on your iPhone

You can send audios, photos, and video messages in the Message app ▢ using SMS/MMS or iMessage service. You can also save, share, or print attachments.

Sending an audio message

- To record an audio message in a conversation, touch and hold ◉.

- Tap ⊙ to playback the audio message before you send it.

- Tap ⬆ to send the audio message or ⊗ to cancel.

N.B. iPhone automatically deletes audio messages two minutes after listening to them to save space unless you tap Keep. To always keep your audio messages, go to **Settings** ⚙ > **Messages** > **Expire** (below Audio Messages), then tap **Never**.

Listening or replying to an audio message

- To play an incoming audio message, raise your iPhone to your ear.
- Raise it again to reply.

If you want to turn this feature on or off, go to **Settings** ⚙ > **Messages** > then turn off **Raise to Listen**.

Sending a photo or video

1. In Messages, you can do any of the following while typing a message:
   - **Taking a photo within Messages**: Tap 📷, frame the shot in viewfinder > tap ⭕.
   - **Taking a video within Messages**: Tap 📷, select a Video mode > tap ⚫.
   - **Choosing an existing photo or video**: Tap 🖼 in the app drawer, then swipe left to go through recent shots or tap All Photos.

Tap here to choose
an existing
photo or video

2. Tap ⬆ to send the message or ✕ to cancel.

Marking up or editing a photo

Before you send a photo in a Message conversation, you can mark up or edit the photo.

Mark up the photo
before sending it
in a message
conversation

1. Tap 🖼 in the app drawer, and pick a photo.

2. Tap the photo in the message bubble, and do either of the following:

   - Tap **Markup**, then use the Markup tools to draw on the photo > **Save**.

   - Tap **Edit**, then use the photo editing tools to edit the photo > **Done**.

3. Tap Done > add a message, and Tap ⬆ to send the photo, or tap ⓧ to erase the photo from the message bubble.

Adding camera effects to a photo or video

After taking a photo or video in a Messages conversation, you can add cool camera effects, such as stickers, text labels, and shapes.

1. In a Messages conversation, Tap 📷, then choose either Photo or Video mode.

2. Tap ⭐, then do any of the following:

   - Tap 😀, then choose a Memoji to add to your photo.

   - Tap 🔵, and choose a filter to apply to your photo.

   - Tap (Aa) to add a text label or tap 〰 to add a shape.

   - Tap 😀 to add a Memoji sticker, or tap 😎 to add an Emoji sticker.

3. Tap ✕ to close the effect window.

4. Tap ○ to take your photo or tap ● to record your video.

5. Click on the icon ⬆ to send the photo or video directly or tap Done to add the video or photo to the message bubble where you can add a text message.

Sending animated effects in Messages on iPhone

You can animate a single message with a bubble effect or use a full-screen effect (for example, confetti or balloons) in the Message app 💬. You can also send a private message with invisible ink that remains blurred until the recipient reveals it by swiping it.

Sending handwritten messages

You can send handwritten messages by using your fingers to write a message. The recipient sees your message animate, just as ink flows on paper.

- Rotate your iPhone to landscape orientation while in a conversation.
- Tap 𝒪 on your keyboard.
- Use your fingers to write a message, or choose a saved message at the bottom > Done.
- Tap ⬆ to send your handwritten message or tap ⊗ to cancel.

Tap to type a

longer message

Choose a saved message.

Touch and hold

to delete it.

Tap to return

to the keyboard

Using bubble effects in Messages on iPhone

This feature allows you to animate the message bubble with bubble effects.

- In a conversation, you can insert a photo or Memoji or write a message.

- Touch and hold ⬆ and click on the gray dots that appear to preview different bubble effects.

- Tap ⬆ to send your message or Click on this ⊗ icon to cancel.

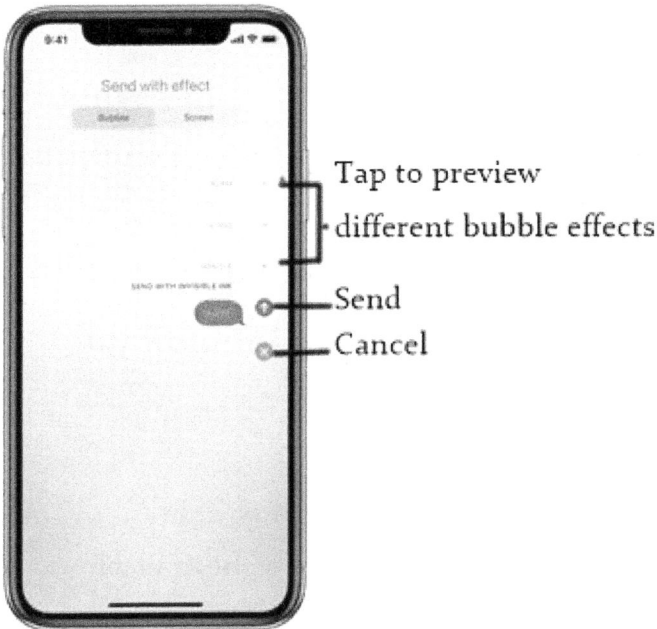

Using full-screen effects in Messages on iPhone

This feature allows you to animate your message screen with full-screen effects.

- In a conversation, write a message or insert a photo or Memoji.

- Touch and hold ⬆, then tap the screen.

- Swipe left or right to preview the available screen effects.

- Tap ⬆ to send your message or click on this ⊗ icon to cancel.

N.B. Messages automatically the following full-screen effects for specific text strings:

- Confetti for "Congratulations"
- Balloons for "Happy birthday."
- Fireworks for "Happy New Year"

Using Memoji in Messages on iPhone

In the Message app ⬜, you can use Memoji and personalized Memoji sticker packs that match your mood and personality to express yourself better in a conversation. You can also send animated Memoji messages that mirror your facial expressions and record your voice.

Creating your own Memoji

You can design your own personalized Memoji. Choose facial features, skin color and freckles, hairstyle and color, glasses, headwear, etc.

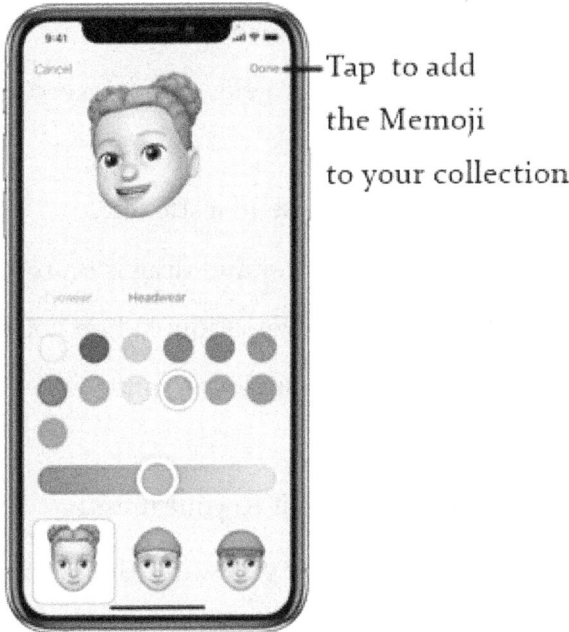

Tap to add the Memoji to your collection

- While in a conversation, tap 😃, then tap +.
- Tap each feature and choose your preferred options. Your character comes to life as you add features to your Memoji.
- To add the Memoji to your collection, tap **Done**.

To delete, edit or duplicate a Memoji, tap 😃, select the Memoji, then tap ⋯.

Sending Memoji and Memoji Stickers

Sticker packs based on your Memoji and Memoji characters are automatically generated by Messages. Stickers enable you to express a range of emotions in new ways.

1. While in a conversation, tap this 😀 icon.

2. Tap a Memoji in the top row to view the stickers in the sticker pack.

3. Do either of the following to send a sticker:

    •   Touch and hold a sticker and drag it on top of a message in your conversation. When you add it to the message, the sticker is sent automatically.

    •   Tap the sticker to add it to your message bubble. Attach a text if you want, then tap ⬆.

Sending animated Memoji or Memoji recordings

You can send animated Memoji messages that mirror your facial expressions and record your voice.

Tap to record
your voice and
facial expressions

- While in a conversation, tap , then choose a Memoji.

- Tap to record your voice and facial expressions. Tap the red square to stop your recording. Tap Replay to review your message.

- Tap to send your message or click on this icon to cancel.

Sending and receiving text messages on iPhone

Send and receive texts, photos, videos, and audio messages with the Message app ⬭. You can also use animated effects, Memoji stickers, iMessage apps to personalize your messages.

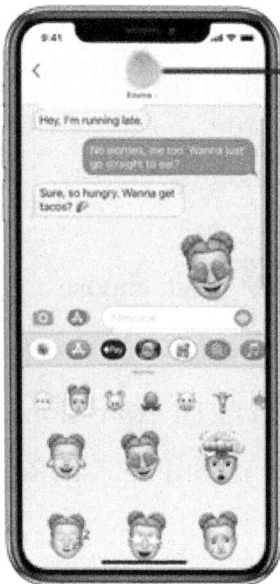

Tap view conversation details and contact info

Sending a message

Do the following to send a text message to one or more people:

1. At the top of your screen, tap ☑ to start a new message or tap an existing message.

Tap to start
a new message

Unread messages
are indicated
by a blue dot

2. Enter the contact name, phone number, or Apple ID
    of each recipient, or you can tap ⊕ and select one or
    more contacts.

    •   On models with dual sim, to send an
        SMS/MMS message with a different line, tap
        the default line that is shown, then choose the
        other line.

3. Tap the next field > type your text message, then tap
    ⬆ to send.

    •   If a message can't be sent, an alert ⊕ appears.
        Tap the alert to resend the message.

◉ **TIP:** Drag the message bubble to the left to see what
time a message was sent or received.

Click on the name or phone number at the top of your
screen, then tap ⓘ to view conversation details. Swipe from
the left edge or tap ‹ to return to the Messages list from a
conversation.

Replying to a message

⏺️**Ask Siri.** Say something like:

- "Reply, that's great news!"
- "Send a message to Eve saying how about tomorrow."
- "Read my last message from Matt."

💡 **TIP:** When you have AirPods Pro, AirPods (2nd generation), or other supported headphones on, Siri can read your incoming messages, and you can reply by speaking the message you want Siri to send.

Or do the following:

1. On your Message list, tap the conversation you intend to reply.
2. Tap the text field and type your message.
3. Tap ⬆️ to send.

💡 **TIP**: Use a Tapback expression (e.g., a heart or a thumbs up) to quickly reply to a message. Double-tap on the message bubble that you want to reply to and select a Tapback.

Sharing your name and photo

Once you start or respond to a new message in Messages, you can share your name and photo. Your photo can be a

Memoji, or you can make a custom image. When you're accessing Messages for the first time, follow the prompts on your iPhone to choose your name and photo.

1. If you want to change your name, photo, or sharing options, open **Messages**, tap ⋯ , tap **Edit Name and Photo,** then do any of the following:

    -   **Change your name**: Tap the text field where the name you intend to change appears.

    -   **Change your profile image**: Tap **Edit**, then select an option.

    -   **Turn sharing on or off**: Toggle the switch next to **Name and Photo Sharing** (green indicates that it is on).

    -   **Change who can see your profile**: Select an option below **Share Automatically** (**Name and Photo Sharing** must be turned on).

N.B. You can also use your message's name and photo for your Apple ID and My Card in Contacts.

Pin a conversation

People you contact most always comes first in the list when you pin specific conversations to the top of the Messages list.

To pin a conversation, you either:

- Swipe right on a conversation, then tap ⊼.
- Or touch and hold a conversation, then drag it to the top of the Messages list.

Swipe right on a conversation to pin it

Unpin a conversation

Unpin specific conversations at the top of your Messages list.

To unpin a conversation, you either:

- Touch and hold a conversation, then tap ⊼.
- Or touch and hold a conversation, then drag the message to the bottom of your Messages list.

Switching from a Message conversation to an audio call or FaceTime

You can initiate an audio call or FaceTime when you are chatting with a person in Messages.

- Click on the profile picture or the name at the top of the conversation (Messages conversation).

- Then tap audio or FaceTime.

Sending and receiving money with Apple Pay on iPhone (for U.S. only)

In the Message app , make use of Apple Pay to send and receive money quickly and easily. You don't have to download any other app, and you can use the cards you already have with Apple Pay.

Tap to preset this amount.

Sending a payment in Messages

🎙️**Ask Siri.** Say something like: "Apple Pay 60 dollars to Matt for the zipline tour" or "Send 10 dollars to Eve for lunch with Apple Pay."

Or do this:

- While in an iMessage conversation, tap 🍎Pay > enter an amount.

- Tap **Pay**, attach a comment if you want > ⬆️.

- Review the payment information. You can use your debit card in Wallet to pay the balance if you have insufficient funds in Apple Cash

- Use Face ID, Touch ID, or your passcode to authenticate the payment.

- You can only cancel a payment when it has not been accepted. To cancel, tap the payment bubble, then tap **Cancel Payment**.

Requesting a payment

🎙️**Ask Siri.** Say something like: "Ask Eve for the 34 dollars with Apple Pay."

Or do this:

- While in an iMessage conversation, tap 🍎Pay.

- Enter the amount > **Request**.

Changing message notifications on iPhone

You can set up and manage your message notifications from the Messages app 💬 and filter unknown senders.

Managing notifications for your messages

1. Go to **Settings** ⚙️ > **Notifications** > **Messages**.

2. You can do the following:

   - Turn on or off **Allow Notifications**.

   - Set the locations and position of message notifications.

- Choose the alert sound for your message notifications.
- Choose when you want message previews to appear.

Set the alert sound for your messages

1. Go to **Settings** ⚙ > **Sound & Haptics** (on supported models) or **Settings** ⚙ > **Sound.**
2. Tap Text Tone, then do any of the following:
   - Tap **Vibration** and choose your preferred option.
   - Below **Alert Tones**, choose your preferred sound.
   - Tap **Tone Store** to download additional alert sounds from the **iTunes Store**.

Assigning a different ringtone to a contact

- Go to **Contacts**, then select a contact.
- Tap **Edit** > **Text Tone.**
- Below **Alert Tones**, choose your preferred option.

Turn on Emergency Bypass to allow alerts for messages sent by this contact even when Do Not Disturb is on.

Muting notifications for a conversation

- While in the **Messages list**, swipe left on the conversation.
- Then tap **Hide Alerts**.

Block, report, and filter messages on iPhone

The Message app ⬜ allows you to block unwanted messages, report junk or spam messages, and filter messages from unknown senders.

Blocking messages from a specific number or person

- While in a Message conversation, tap the number or name at the top of the conversation, then tap ⓘ at the top right of your screen.
- Tap **info**, scroll down, then tap **Block this Caller.**

To view and manage all your phone numbers and blocked contacts, go to **Settings** ⚙ > **Messages** > **Blocked Contacts**.

Reporting junk or spam messages

With iMessage, if you receive a message from someone who isn't saved in your contacts, it might be identified as possible junk or spam.

Tap to report junk or spam messages

- In the message, tap **Report Junk** > **Delete** and **Report Junk**.

N.B. Reporting junk or spam doesn't prevent the sender from sending more messages. The better option is to block the number to stop receiving them.

Contact your carrier to report spam or junk messages you receive with SMS/MMS.

Filter iMessage messages from unknown senders

Turn on Filter Unknown Senders to stop receiving iMessage notifications from senders who aren't in your contact, and to also sort the messages into the Unknown Senders tab in the Messages list.

- Go to **Settings** ⊚ > **Messages** > **Filter Unknown Senders**.

Deleting messages on iPhone

You can delete messages and entire conversations in the Messages app ⬛. With Messages on iCloud, anything you delete from iPhone is also deleted from your other Apple devices that are signed in with the same Apple ID if Messages in iCloud is turned on.

Deleting a message

- Touch and hold the message bubble, then tap More.
- Select the message bubble you wish to delete > 🗑.

Deleting a conversation

- While in the **Messages list**, swipe left on the conversation.
- Then tap **Delete**.

8: Calendar

The iPhone's Calendar app 15 provides a great way to always be on top of your routine.

Creating and editing events in Calendar on iPhone

Create and edit events, meetings, and appointments with the Calendar app 15

Tap to view list of events

Tap to change calendars or accounts

View your invitations

⬤**Ask Siri.** Say something like:

- "Set up a meeting with Matt at 12."
- "Do I have a meeting at 2:30?"
- "Where is my 2:30 meeting?"

Adding an event

- In day view, tap ✛ at the top left of your screen.
- Fill in the details of the event > **Add.**

Adding an alert

Set an alert for your event to be reminded of the event beforehand.

- Tap the event, then tap Edit at the top right of your screen.
- Tap alerts in the event details.
- Then choose your preferred time to be reminded (for example, 5 minutes before," "At the time of the event," etc.).

Find events in other apps

If you turn on Siri Suggestions in App, Siri suggests events found in other apps, such as flight reservations and hotel bookings, etc., so that you can add them to Calendar easily.

- Go to **Settings** ⚙ > **Calendar** > **Siri & Search.**

- Then turn on **Siri Suggestions in App**.

Turn on **Learn from this App** for Calendar to allow Siri to make suggestions in other apps based on how you use Calendar.

Editing an event

After adding an event, you can change the time of the event and any other event details.

- **Changing the time**: In day view, touch and hold the event, then set it to a new time or adjust the grab points.
- **Changing event details**: Tap the event > **Edit**, then at the details of the event, tap a setting to change it, or tap in a field to write new information.

Deleting an event

- In day view, tap the event.
- Then at the bottom of your screen, tap **Delete Event**.

Sending and receiving invitations in Calendar on iPhone

With the iPhone's Calendar app 15 , you can send and receive event and meeting invitations.

N.B. Calendar servers like iCloud, Microsoft Exchange, and some CalDAV servers let you send and receive meeting invitations. Not all calendar servers support every feature.

Inviting others to an event

- Tap the event > **Edit** > **Invitees** > **Add invitees**.
  Or if the event was scheduled by someone else, tap the event > **Invitees**, then tap ✉.
- Type in the names or email addresses of invitees, or tap ⊕ to select from Contacts.
- Then tap **Done** or **Send** (if the event was scheduled by someone else).

Replying to an event invitation

- Tap an event notification to respond to it. Or in **Calendar**, tap **Inbox**, then tap an invitation.
- Then choose a response (Accept, Maybe, or Decline). To respond to invitations received via email, tap the underlined text > **Show in Calendar**.

Suggesting a different meeting time

You can propose a different time for a meeting invitation you have received.

- Tap the meeting > **Propose New Time.**

- Tap the time, and enter a new time.

The organizer will receive either an email or a counter-proposal with your suggestion (depending on your server's capabilities).

Sending an email to attendees

- Tap an event that has attendees > **Invitees**, then tap ✉.

Searching for events in Calendar on iPhone

In the Calendar app 🗓, type in title, location, invitees, or notes to search for events.

- From the Calendar app 🗓, tap 🔍, then enter the text you want to find in the search field.

🎤 **Ask Siri.** Say something like: "What is on my calendar for Saturday?"

Customizing your calendar on iPhone

You can customize your calendar by choosing which day of the week Calendar starts with, choosing alternate calendars (for example, to display Hebrew or Chinese dates), displaying week numbers, overriding the automatic time zone, etc.

- Go to **Settings** ⚙ > **Calendar** > then choose your preferred settings and features.

Keeping your Calendar up to date across your devices

You can use iCloud to keep your Calendar information up to date across all your devices signed in with the same Apple ID.

- Go to **Settings** ⚙ > {your name} > **iCloud** and turn on Calendar.

Setting up multiple calendars on iPhone

The Calendar app 📅 allows you to set up multiple calendars to keep track of different events.

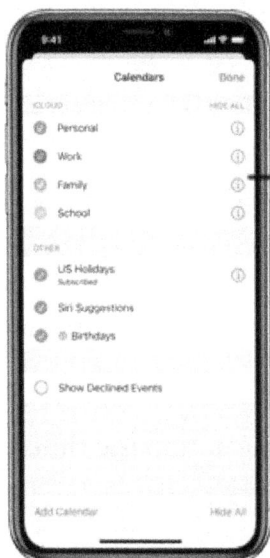

Choose which calendars to view

See multiple calendars at once

Tap Calendars at the bottom of the screen, then do any of the following to view multiple calendars:

- Choose the calendars you want to view.
- Tap US Holidays to add national holidays with your events.
- Tap Birthdays to add birthdays from Contacts with your events.

Setting a default calendar

You can choose one of your calendars as the default calendar. Events added using Siri or other apps will be automatically added to your default calendar.

- Go to **Settings** ⚙ > **Calendar** > **Default Calendar**.
- Then choose the calendar you want to use as your default calendar.

Turn on iCloud, Google, Yahoo, or Exchange

- Go to **Settings** ⚙ > **Calendar** > **Accounts** > **Add Account**.
- Then tap a mail service (for example, iCloud or Microsoft Exchange), then sign in to your account.

- Or tap Other, tap Add CalDAV Account or Add Subscribed Calendar, then enter your server and account details.

Subscribing to a calendar

- Go to **Settings** ⚙ > **Calendar** > **Accounts** > **Add Account** > **Other**.
- Tap **Add Subscribed Calendar**.
- Enter the URL of the .ics file and other required server information to subscribe.

Adding a CalDAV account

- Go to **Settings** ⚙ > **Calendar** > **Accounts** > **Add Account** > **Other**.
- Tap **Add CalDAV** account, then enter your server and account details.

Moving an event to another calendar

- Tap the event > **Calendar**, then select the calendar you want to move the event to.

9: Clock

Keep track of time and set alarms with the Clock app ◯.

- From your apps, tap **Clock**.

World Clock

Use the World Clock to keep track of the local time in different time zones around the globe.

🎙 **Ask Siri.** You can say something like: "What time is it?" or "What time is it in Paris?"

Tap to manage the World Clock

1. From the **Clock app** ⊘, tap **World Clock.**

2. Tap **Edit** and do any of the following to manage your list of cities:

   • **Adding a city**: Tap ＋, then choose a city to add to your list.

   • **Deleting a city**: Tap ⊖ to delete a city.

   • **Reordering the cities**: Move ≡ up or down.

Setting an alarm on iPhone

Set alarms for specific times with the Clock app ⊘.

🎙**Ask Siri.** Say something like "Set an alarm for 6:30 a.m. every Monday" or "Wake me up tomorrow at 5 a.m."

Tap to change
or delete the alarm ——

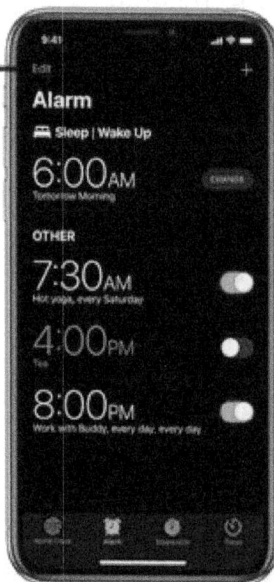

Set an alarm

1. From the **Clock app** ⊙, tap **Alarm**, then Tap ＋.

2. Set the time, and then you can choose from the following options:

    *   **Repeat**: Choose the days of the week you want the alarm to repeat.

    *   **Label**: Name the alarm to remember what it was for, like "Feed the goldfish."

    *   **Sound**: choose your preferred tone or vibration.

    *   **Snooze**: Have a few more minutes of sleep.

3. Tap **Save**. Tap **Edit** to change or delete the alarm.

Making changes to your next Wake Up alarm

You can use the Health app to set up a sleep schedule that includes Bedtime and Wake Up times, Wind Down options, alarm settings, etc., for each day of the week.

When you have set up a sleep schedule in the Health app, you can use the Clock app to make changes to your next Wake Up alarm.

*   Tap **Alarm** > **Change**.

*   Drag 🛏 and 🔔 to adjust your sleep and wake times.

*   Choose alarm options, such as alarm sound, snooze.

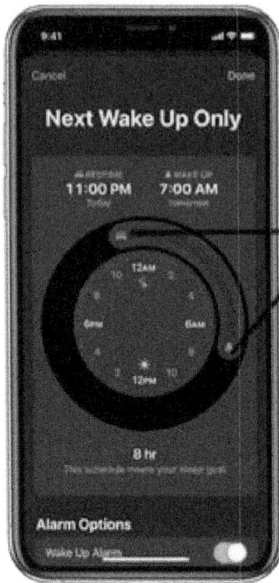

Drag to adjust your sleep and wake times

Using the stopwatch or timer on iPhone

In the Clock app ⊘, you can measure the duration of an event with the stopwatch, and you can also use the timer to count down from a specified time.

🎤 **Ask Siri.** Say something like: "Set the time for 45 seconds" or "Stop the timer."

Setting the timer

- From the **Clock app** ⊘, tap **Timer** > Set the time duration and a sound to play when the timer stops.
- Then tap **Start**.

The timer continues even if your iPhone goes to sleep or if you open another app.

Using the stopwatch to track time

- Tap **Stopwatch**. Swipe the stopwatch to switch between the digital and analog faces.
- Then tap **Start**. The timing continues even if your iPhone goes to sleep or if you open another app.
- Tap **Lap** to record a lap or split
- Tap **Stop** to record the stopwatch's final time.
- Tap **Reset** to clear the stopwatch.

10: Calculator

The Calculator app ▦ features both the standard and scientific calculator.

🎙️**Ask Siri.** Say something like: "What is 56 times 12?" or "What is 13 percent of 765?"

To switch to the scientific calculator, rotate your iPhone to landscape orientation

Using the scientific calculator

To switch to the scientific calculator, rotate your iPhone to landscape orientation.

All Clear key

Deleting, clearing, or copying a calculation

- **Deleting the last digit**: If you mistakenly input a number, swipe left or right on the display at the top.

- **Clearing the display**: To delete the last entry, tap the Clear (C) key, or tap the All Clear (AC) key to clear the display.

- **Copying a calculation result**: Touch and hold a calculation result, tap Copy, then paste the result somewhere else.

11: The iPhone Health app

The Health app ♥ is designed to look after your wellbeing and lead a healthier lifestyle. It does this by tracking how you sleep, what you eat, and other things that contribute to a healthier lifestyle.

Collecting health and fitness data in Health

You can use the Health app ♥ to track your daily footsteps and the flights of stairs you climb. You can manually add other data like your caffeine intake and body weight. And also, use other apps (such as fitness and nutrition apps) and devices that are compatible with Health (such as Apple Watch, AirPods, blood pressure monitors, and weight scales) to track additional data.

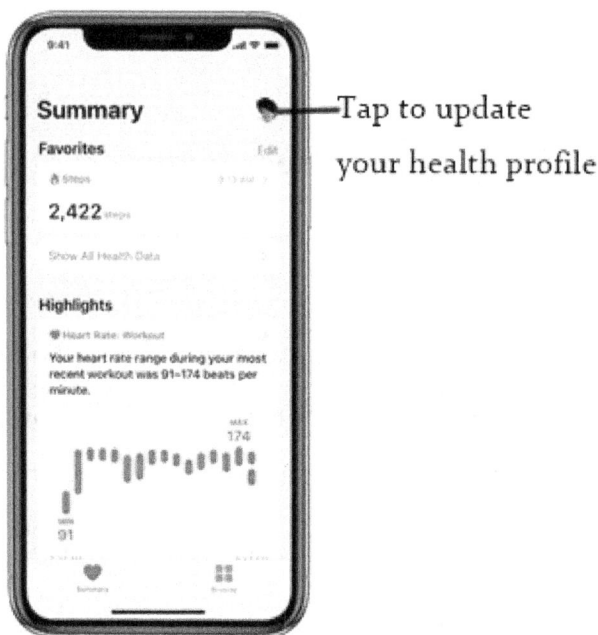

Tap to update your health profile

Manually updating your health profile

When you open the Health app 💛 for the first time, you will be asked to set up a health profile, which entails basic information about yourself, such as your date of birth and sex. You could update your health profile later if you didn't supply all of the requested information.

- Tap your initials or profile picture at the top right.

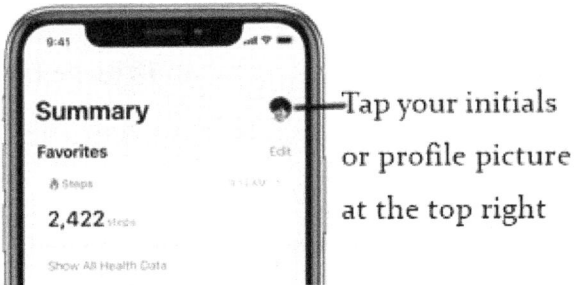

Tap your initials
or profile picture
at the top right

- Tap **Health Details** > **Edit**.

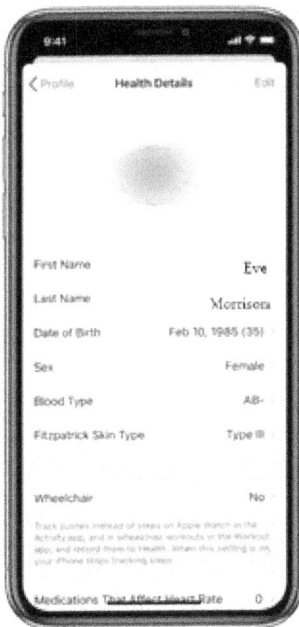

- Enter a field, make your changes > **Done**.

Manually adding data to a health category

1. At the bottom right of your screen, tap **Browse** to
    display the Health Categories screen, then do the
    following:

    - Tap a category, and scroll down to see all
      categories.

    - Tap the search field and enter the name of a
      category (such as body measurements) or
      specific data (such as weight).

    - To update the data, tap $>$.

    - At the top-right corner of your screen, tap **Add
      Data**.

    - Add your information, then at the top-right
      corner of your screen, tap **Add** or **Done**.

Collecting data from other devices and apps

- **From Apple Watch**: After pairing your iPhone to
  Apple Watch (see Accessories), Apple Watch
  automatically sends a periodic heart rate
  measurement to Health. You can also set up Apple
  Watch to send noise levels, activity metrics, etc., to
  Health.

- **From headphones:** After connecting Airpods,
  EarPods, and other compatible headphones to your

iPhone, Health automatically receives the headphones' audio levels.

- **Apps downloaded from App Store**: When setting up the app, you can give permission to share data with Health.

Viewing health and fitness information

The Health app enables you to view your health and fitness information in one place. For example, you can track your symptoms to see whether they are improving or not and also see how well you're meeting the goals for sleep, activity, mindfulness, etc.

Viewing your highlights

At the lower left of your screen, tap Summary, then scroll down to see recent highlights of your health and fitness data.

Tap > to see more details about a category.

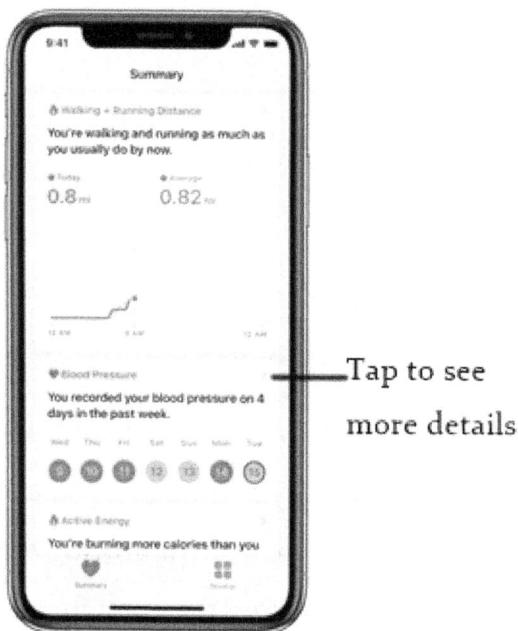

Tap to see
more details

Adding or removing a health category from favorites on the Summary screen

- At the lower left of your screen, tap **Summary**, tap Edit for the Favorites section, click on a category to turn it on or off > **Done**.

Viewing details in the health categories

1. At the bottom right of your screen, tap **Browse** to display the Health Categories screen, then do any of the following:
   - Tap a category, and scroll down to see all categories.

- Tap the search field Tap the search field, then enter the name of a category (such as Nutrition) or a specific data (such as carbohydrates).

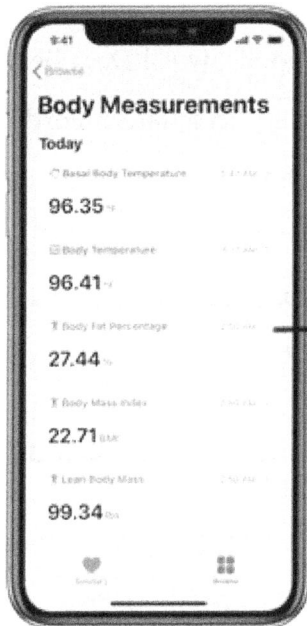

Tap to see more details

2. Tap $>$ to see details about any of the data. Depending on the type of data, you can do the following:

- **Seeing yearly, monthly, weekly views of the data**: Tap the tabs at the top of your screen.
- **Manually entering data**: At the top-right corner of your screen, tap Add Data.

- **Moving a data type to favorites on the Summary screen**: Turn on Add to Favorites.

- **Viewing which devices and apps are allowed to share data:** Tap Data Sources & Access below Options.

- **Delete data**: Tap Show All Data below Options, swipe left on a data record > Delete. If you want to delete all data, tap Edit > Delete all.

- **Changing the measurement unit**: Tap Unit below Options, then choose a different unit.

Tracking your menstrual cycle in Health

You can use the Health app 🫀 to track your menstrual cycle to get period and fertility window predictions.

Getting started with cycle tracking

- At the bottom right of your screen, tap **Browse** > **Cycle Tracking**.

- Tap Get Started, then follow the prompts.

Insert the requested information about your last period to help improve predictions for your period and fertility window.

Log your cycle information

1. At the bottom right of your screen, tap **Browse** >
   Cycle Tracking.

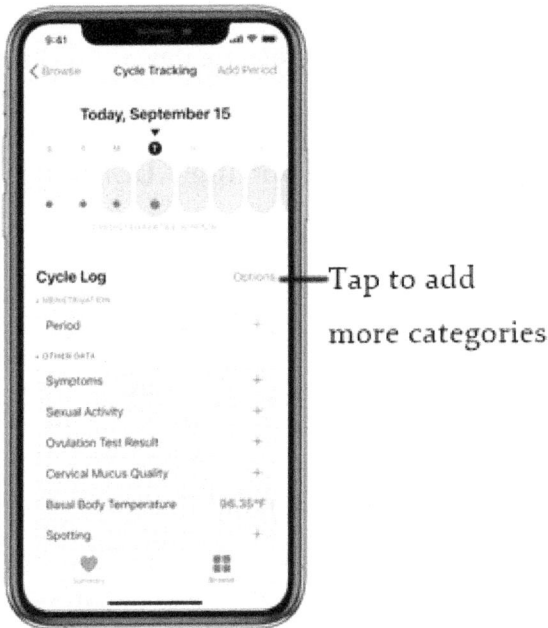

Tap to add more categories

2. Then do any of the following:
   - **Log a day period**: In the timeline on top of
     your screen, tap a day. Tap **Period** below
     Cycle Log to log the flow level for that day,
     then choose an option. Logged days are
     indicated on the timeline with solid red
     circles. Tap a logged day to remove it.
   - **Log symptoms**: To select a day, drag the
     timeline at the top of your screen, tap

Symptoms, then select all symptoms that apply. When through, tap **Done**. Days with symptoms are indicated on the timeline with purple dots.

- **Log spotting**: To select a day, drag the timeline at the top of your screen, tap **Spotting** > **Had Spotting** > **Done**.

3. If you want to add additional categories, e.g., basal body temperature and ovulation test results, **tap Options**, then choose the categories.

View the cycle timeline

At the bottom right of your screen, tap **Browse** > **Cycle Tracking**. Timeline information is represented in the following format:

- **Solid red circle**: For days you logged for your period.
- **Purple dots**: For days you logged for having symptoms.
- **Light red circles**: For your period prediction. If you want to hide or show predicted period days, tap **Options**, then turn **Period Prediction** off or on.
- **Light blue days**: For the prediction of your likely fertility window. Do not use fertility window prediction as a form of contraception. If you want to

hide or show fertility window prediction, tap
Options, then turn **Fertility Prediction** off or on.

Drag the timeline to select different days. Data that you
logged for the selected day is shown below in the Cycle Log.

**Changing period and fertility notifications and other
cycle tracking options**

- At the bottom right of your screen, tap **Browse** >
  Cycle Tracking.
- Scroll down and tap **Options**.
- Tap an option to turn it on or off.

View your cycle history and statistics

- At the bottom right of your screen, tap **Browse** >
  Cycle Tracking.
- Scroll down to see timelines of your three most
  recent periods. To see related statistics, scroll further.
- To see older information and more details for Cycle
  History or Statistics, tap $>$ in that section of the
  screen.

Setting up a sleep schedule

The Health app 💤 can help you reach your sleep goals by
scheduling times for relaxing at the close of your day,

getting to bed, and waking up. You can create multiple schedules, e.g., one for weekends and a second one for weekdays. For both schedules, you can set up how you want your iPhone to wake you up by choosing an alarm and other options.

Enable Sleep Mode for iPhone to protect your rest and reduce distractions. Sleep mode turns on Do Not Disturb and simplifies your Lock Screen. You can set up a Sleep Schedule to begin before it is bedtime.

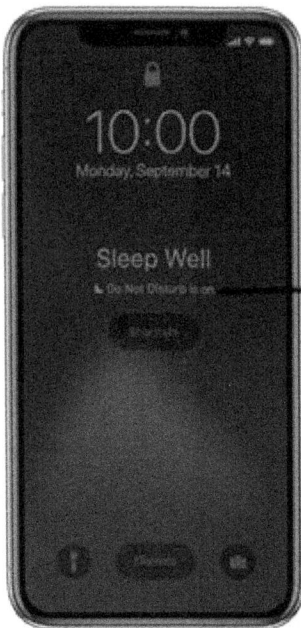

Sleep mode turns on Do Not Disturb and simplifies your Lock Screen

Add Unwind Down Shortcuts to access these activities on the Lock Screen (see the picture below) if you enjoy relaxing before sleep by using the iPhone for activities like listening to music or reading.

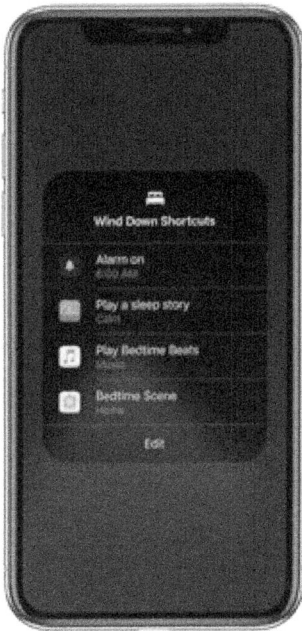

—Add Unwind Down Shortcuts to access these activities on the Lock Screen

Set up your first schedule

In the Health app 💗, you set up a schedule for winding down, going to bed, and waking up for one or more days of the week.

- At the bottom right of your screen, tap **Browse** > **Sleep**.
- Swipe up, and below Set Up Sleep, tap **Get Started**.
- Then follow the prompts.

Change your next alarm

You are allowed to make temporary changes to a schedule.

- At the bottom right of your screen, tap **Browse** > **Sleep**.
- Scroll down to Your Schedule, then below Next, tap **Edit**.
- Drag 🛏 and 🔔 to adjust your sleep and wake times.
- Choose alarm options, such as alarm sound, snooze.
- Then tap **Done**.

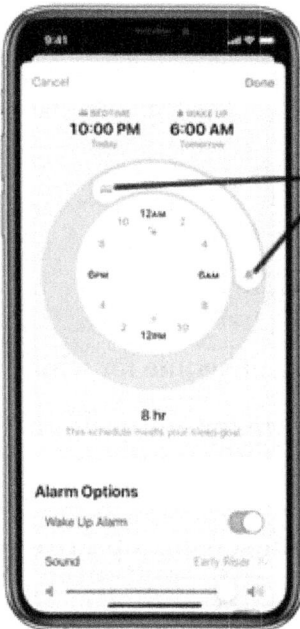

Drag to adjust your sleep and wake times

N.B. You can use the Clock app 🕐 to change your next wake-up alarm (*see* Setting an alarm on iPhone (page 180)). Your usual schedule resumes after your next wake-up alarm.

Changing or adding a sleep schedule

1. At the bottom right of your screen, tap **Browse** > **Sleep**.

2. Scroll down to Your Schedule, and tap **Full Schedule & Options.**

3. Then do the following:

   - **Changing a sleep schedule**: Tap **Edit** for the sleep schedule you want to change.

   - **Adding a sleep schedule**: Tap Add **Schedule for Other Days.**

4. Do the following:

   - **Set the days for your schedule**: To add or remove a day from your schedule, tap the day at the top of the screen. The schedule only applies to days shown with solid-color circles.

   - **Adjust your bedtime and wake-up schedule**: Drag 🛏 and 🔔.

   - **Set the alarm options**: Turn on or off **Wake Up Alarm**.

   - **Cancel or remove a sleep schedule**: To remove an existing schedule, tap **Delete Schedule** (at the bottom of the screen) or tap

Cancel (at the top of the screen) to cancel
when creating a new one.

5. When through, tap **Done or Add**.

To turn off all your sleep schedules, At the bottom right
of your screen, tap **Browse** > **Sleep** > **Full Schedule &**
Options, then turn off **Sleep Schedule** (at the top of the
screen).

Changing your Wind Down schedule and activities

*   At the bottom right of your screen, tap **Browse** >
    Sleep > **Full Schedule & Options**, then do the
    following:

    o **Change Sleep Schedule to begin before it is**
       bedtime: Tap Wind Down, then select a time.

    o **Adding or removing activities for winding**
       down: Tap **Wind Down Shortcuts** > **Add**
       Another Shortcut or ⊖ to remove. When
       iPhone is in Sleep Mode, Wind Down
       Shortcuts for activities like listening to music
       or reading will appear on the Lock Screen.

Changing your sleep goal and other options

*   At the bottom right of your screen, tap **Browse** >
    Sleep > **Full Schedule & Options**.

- To change your sleep goal, tap **Sleep Goal**, then select a new time.

- Tap **Options** to turn other options on or off.

Viewing your sleep history

Viewing your sleep data in Health provides insight into your sleep habits.

1. At the bottom right of your screen, tap **Browse** > **Sleep.**

2. Then do any of the following:

   - **Viewing sleep data by week or month**: Tap a tab at the top of the screen.

   - **Changing the time span shown in the graph**: Swipe the graph right or left.

   - **Viewing the details for a day**: Tap the column for the day,

   - **Manually adding sleep data**: In the top-right corner of your screen, tap **Add data**.

   - **Getting cumulative sleep data**: Tap **Show More Sleep Data.**

Tap to manually add sleep data

Tap to get cumulative sleep data

Sharing health and fitness data in Health

If you want to share health and fitness data in the Health app ♥, you will have to give permission to other apps. For instance, if you install a workout app, its exercise data can be seen in the Health app.

Controlling the sharing of data among apps and devices

- Tap your initials or profile picture at the top right.

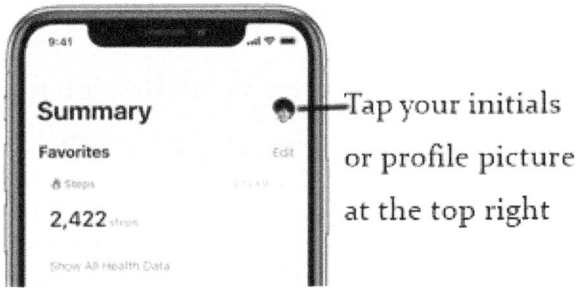

Tap your initials or profile picture at the top right

If you don't see your initials or profile picture, tap Summary or Browse at the bottom of the screen, then scroll to the top of the screen.

- Tap Apps or Devices below Privacy. The items that requested access to Health data are listed on the screen.

- To change the access for an item, tap the item, then turn on or off permission to read data or write data from Health.

Exporting and sharing your health data

- Tap your initials or profile picture at the top right.

    If you don't see your initials or profile picture, tap Summary or Browse at the bottom of the screen, then scroll to the top of the screen.

- Tap **Export all health data**, then choose a sharing method for your data.

Downloading health records in Health

The Health app 💟 gives you access to information from supported health organizations about your conditions, allergies, medications, etc. (not available in all countries or regions).

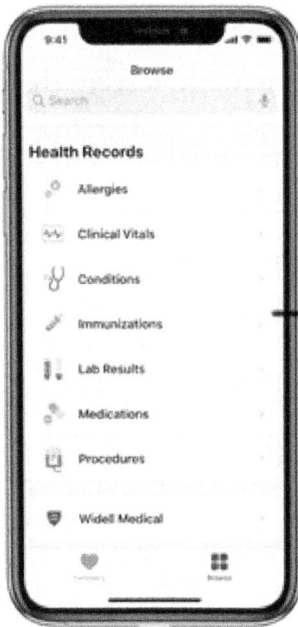

Download your
health records in Health

Setting up automatic downloads

1.   Tap your initials or profile picture at the top right.

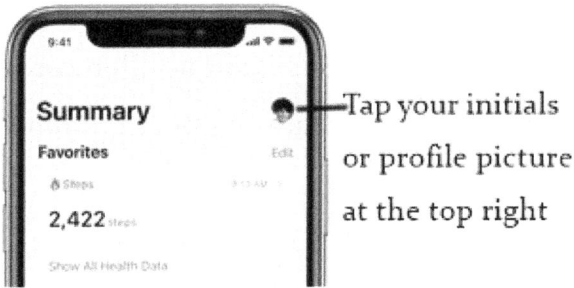

Tap your initials
or profile picture
at the top right

If you don't see your initials or profile picture, tap Summary or Browse at the bottom of the screen, then scroll to the top of the screen.

2. Tap **Health Records**, then do one of the following:

    •   Set up downloads for additional accounts: Tap **Add Account**.

    •   Set up your first download: Tap **Get Started**.

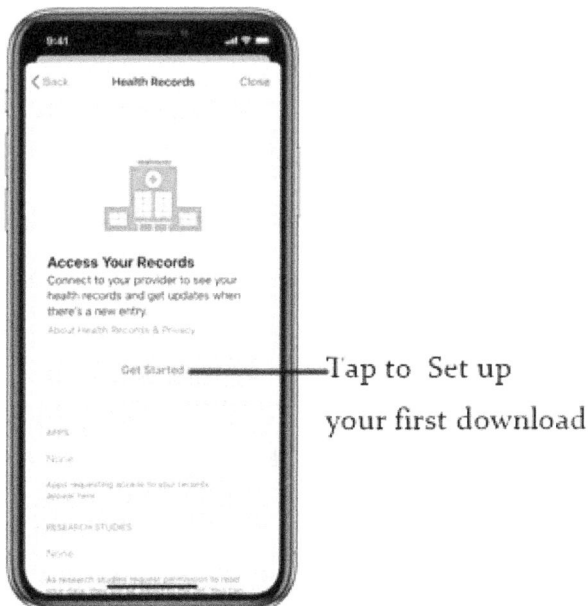

Tap to Set up
your first download

3. Input the name of an organization, such as a hospital or clinic, where you obtain your health records. Or input the name of the state or city where you live to find a list of nearby organizations.

4. Tap a result to open it.

5. Tap the **Connect to Account** button below 'Available to Connect' to go to the sign-in screen for your patient portal.

6. Input the username and password you use for the patient web portal of that organization, then follow the prompts.

Viewing your health records

1. At the bottom right of your screen, tap **Browse** to display the Health Categories screen, then do one of the following:

   • Tap the search field, then input the name of a health record category (e.g., Clinical Vitals) or a type of data (e.g., Blood Pressure).

   • Scroll down, then below Health Records, tap a category (e.g., Clinical Vitals or Allergies).

   • Scroll down, then tap the name of the organization.

To see more details, tap $>$ in that section of the screen.

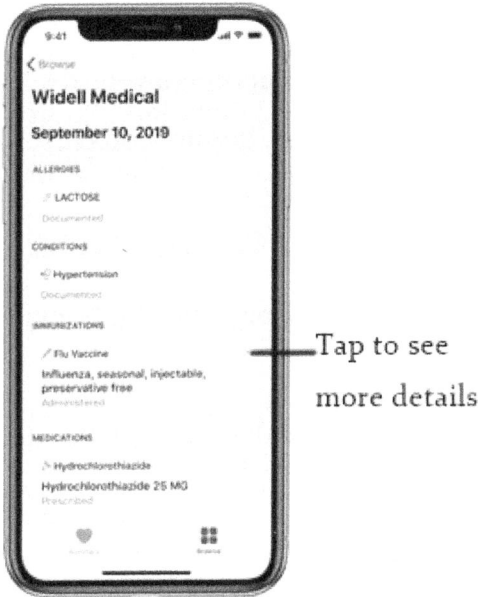

Tap to see more details

Customizing notification settings for health records

- Go to **Settings** ⚙ > Notifications > Health, then choose your preferred options.

Deleting an organization and its records

- Tap your initials or profile picture at the top right.

    If you don't see your initials or profile picture, tap Summary or Browse at the bottom of the screen, then scroll to the top of the screen.

- Tap **Health Records**, then tap the name of an
  organization > **Remove Account**.

Sharing your records with other apps

Third-party apps can demand access to your health records.
Before granting the app access, be sure you trust it with your
records.

- To grant access, choose categories to share (e.g.,
  Allergies, medications, or immunizations) when you
  are asked.
- Choose whether to grant access to both your current
  and future health records or only to your current
  records. If you choose to grant access to only your
  current records, you'd be asked to grant access every
  time new records are downloaded to your iPhone.

If you want to stop sharing health records with an app,
turn off its permission to read data from Health.

Creating and sharing your Medical ID in Health

In the Health app 💙, you can create an emergency Medical
ID that contains information about your allergies, medical
condition, medications, etc. This critical information can be

viewed by first responders directly on your iPhone, even while it is locked.

You can allow iPhone to send this medical information automatically to emergency services through a secure third-party service any time you call or text 911 or use Emergency SOS (for U.S. only) from your iPhone or Apple Watch (watchOS 6.2.5 or later is required if Apple Watch is connected to Wi-Fi or a cellular network without your iPhone close by).

When emergency contacts are added to your Medical ID, they are automatically alerted when you use Emergency SOS from your iPhone or your Apple Watch.

Creating or making changes to your medical ID

1. Tap your initials or profile picture at the top right.

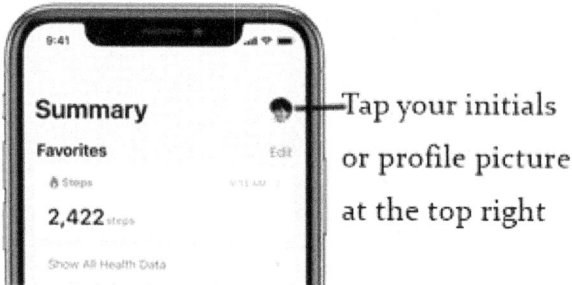

Tap your initials or profile picture at the top right

If you don't see your initials or profile picture, tap Summary or Browse at the bottom of the screen, then scroll to the top of the screen.

2. Tap **Medical ID**, then do either one of the following:

   - **Create a Medical ID**: Tap **Get Started**.
   - **Change your Medical ID**: Tap **Edit**.

Important: Turn on **Share During Emergency Call** to allow iPhone to send your Medical ID information to emergency services whenever you call or text 911 or use Emergency SOS (for U.S. only; 911 is not available in all locations).

Show When Locked is turned on by default to enable first responders to view your Medical ID even when your iPhone is locked. Do not turn this feature off unless you want to prevent responders from viewing your Medical ID.

First responders can view your Medical ID from the Lock Screen by swiping up or pressing the **Home button** (depending on your iPhone), tapping **Emergency** on the Passcode screen > Medical ID.

🎧 **TIP:** To quickly view your Medical ID from the Home Screen, touch and hold the Health app icon > Medical ID.

Managing health features with the Health Checklist

The Health Checklist enables you to review and turn on essential features in the Health app ♡.

1. Tap your initials or profile picture at the top right.

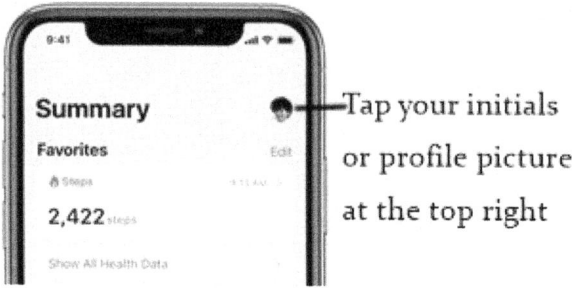

Tap your initials or profile picture at the top right

If you don't see your initials or profile picture, tap Summary or Browse at the bottom of the screen, then scroll to the top of the screen.

2. Tap **Health Checklist**.

3. Tap an item to turn it on or to learn more about it. Tap Back to return the checklist.

4. When through, tap **Done**.

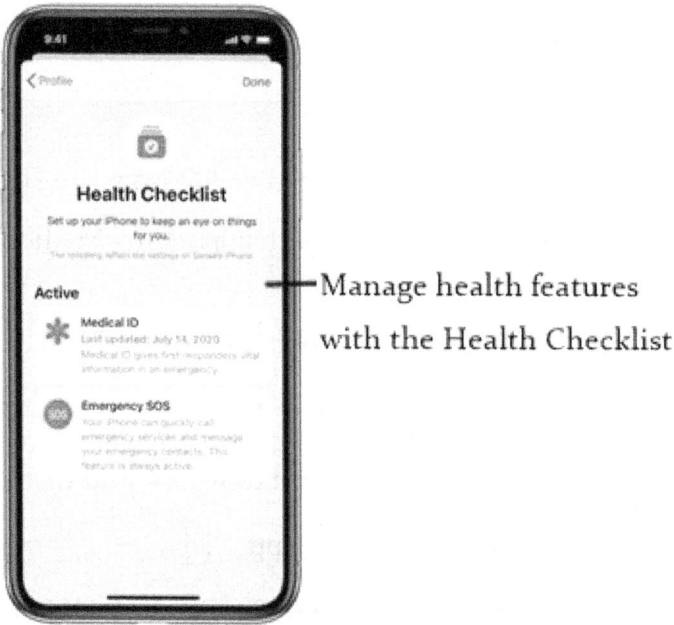

Manage health features with the Health Checklist

Registering as an organ donor in Health (for U.S. only)

In the Health app 🩺, you can register to be an organ or tissue donor with Donate Life America. Your decision to donate is visible to others in your Medical ID. You can remove your registration if you later change your decision.

1. Tap your initials or profile picture at the top right.

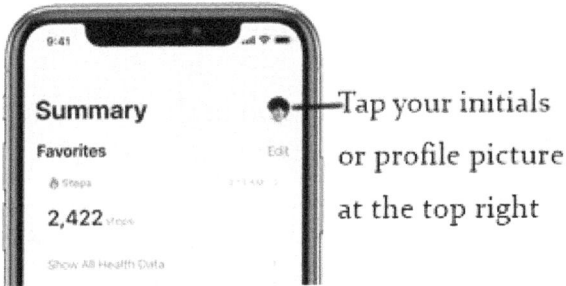

Tap your initials or profile picture at the top right

If you don't see your initials or profile picture, tap Summary or Browse at the bottom of the screen, then scroll to the top of the screen.

2. Tap **Organ Donation**. For an overview of organ donation and Donate Life America, tap **Learn More**.

Registering with Donate Life America

1. Tap your initials or profile picture at the top right.

If you don't see your initials or profile picture, tap Summary or Browse at the bottom of the screen, then scroll to the top of the screen.

2. Tap **Sign Up with Donate Life**.

To remove your registration or change your donor information, tap your initials or profile picture > **Organ Donation** > **Edit Donor Registration**.

Backing up your health data on iPhone

Your health and fitness information in the Health app ♥ is automatically stored in iCloud if you sign in with your Apple ID. iPhone keeps your information encrypted as it goes between iCloud and your device and while stored in iCloud.

Stop storing your Health data in iCloud

1. Go to **Settings** ⚙ > {your name} > **iCloud,** then turn off Health.

Section 3: Blabbing with Siri and iPhone Accessories

12: Learn how to talk to Siri – "Hey, Siri!"

Siri is an intelligent virtual assistant that is unique and exclusive to the iPhone models. Talking to Siri is an easy and super fast way of getting things done. You can ask Siri to set a timer, report on the weather, translate a phrase, find a location, etc. Siri adapts to you, so the more you use Siri, the better it knows your needs.

N.B. To use Siri, your iPhone must be connected to the internet (cellular charges may apply).

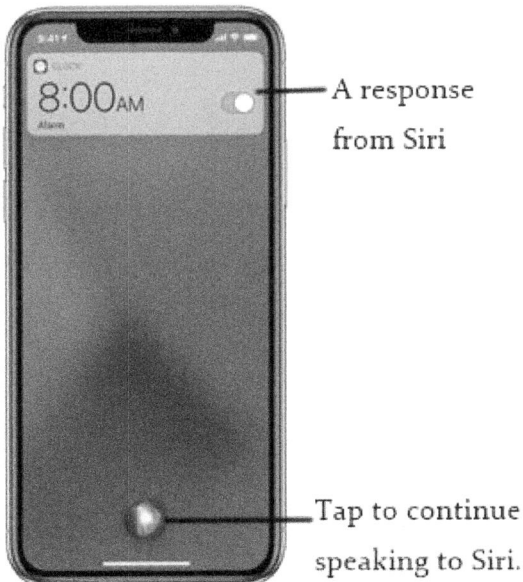

A response from Siri

Tap to continue speaking to Siri.

Getting started with Siri

You can set up Siri when you first set up your iPhone. But if you didn't, go to **Settings** ⚙ > **Siri & Search**, then do either one of the following:

- If you prefer summoning Siri with your voice: Turn on **Listen for "Hey Siri."**

- If you prefer summoning Siri with a button: Turn on **Press Side Button for Siri** (on iPhone models with Face ID) or **Press Home for Siri** (on iPhone models with a Home button).

See Adjusting Siri settings on iPhone (page 228) if you want to change additional Siri settings.

Summoning Siri with your voice

Siri responds out loud when you summon Siri with your voice.

1. To summon Siri, say "Hey Siri," then instruct Siri to perform a task for you or ask a question. For example, say something like "Hey Siri, set an alarm for 5 a.m." or "Hey Siri, what is today's weather report?"

2. If you want to instruct Siri to perform another task or answer another question, say "Hey Siri" again or tap

   🔘.

N.B. If you want to prevent Siri from responding to "Hey Siri," place your iPhone face down, or go to **Settings** 🔘 > **Siri & Search**, then turn off **Listen for "Hey Siri."**

Summoning Siri with a button

When your iPhone is in ring mode, Siri responds out loud when you summon Siri with a button, and when iPhone is in silent mode, Siri responds silently.

1. Depending on your iPhone model, do either one of the following:

   - **iPhone models with Face ID**: Press and hold the side button.

   - **iPhone models with Home button**: Press and hold the Home button.

2. When Siri appears, instruct Siri to perform a task or ask Siri a question.

3. If you want to instruct Siri to perform another task or answer another question, say "Hey Siri" again or tap 🔘.

Correcting your request if Siri misunderstands you

- **Rephrase your request**: Tap 🔘, then say your request differently.

- **Spell out part of your request**: Tap 🌀, then repeat your request by spelling out any words that Siri misunderstands. For example, say "Send a message to," then spell the person's name.

- **Changing your message before sending it**: Say "Change it."

- **Use text to edit your request**: You can edit your request if you see it onscreen. To edit it with text, Tap the request, then use the onscreen keyboard.

Use type to Siri instead of speaking

You can type in your request instead of speaking.

- Go to **Settings** ⚙ > **Accessibility** > **Siri**, then enable **Type to Siri**.

- To make a request, summon Siri and type in what you want Siri to do for you.

Blabbing with Siri

On the iPhone, you can use Siri to perform tasks and get information. Siri allows you to refer to information onscreen (that is, Siri and its response appear on top of what you are currently doing).

- **Getting answers to your questions**: You can use Siri to find information on the web, get arithmetic calculations, get sports scores, etc. Say something like "Hey Siri, what is a supernova," "Hey Siri, what is the formula for calculating the arithmetic mean," or "Hey Siri, what was the final score of the Yankee game tonight?"

- **Using apps to perform tasks on iPhone**: With Siri, you can navigate through apps by using your voice. For example, to set an alarm in Clock, say something like "Hey Siri, set an alarm for 7 a.m. every Monday," or to make a call from Contacts, say something like "Hey Siri, call Eve's mobile."

- **Hear about your day**: You can make Siri tell you about your day, that is, tell you about your weather, news, reminders, calendar events, etc. Say something like, "Hey Siri, what's my update?"

- **Tune in to a radio station**: Say something like "Hey Siri, tune into ESPN Radio," or "Hey Siri, play Wild 94.5."

- **Translate languages**: You can use Siri as a translator. Say something like, "Hey Siri, how do you say Thank You in Mandarin?"

Tap to play the translation again.

- **Send and reply to text messages**: Say something like "Hey Siri, send a message to Eve saying how about tomorrow," or "Hey Siri, reply that's great news." You can also use Siri to send audio messages.

- **Get directions**: Say something like "Hey Siri, find gas stations near me," or "Hey Siri, get directions home."

- **Get more examples**: Although additional examples appear throughout this manual, you can still get more examples from Siri. Say something like, "Hey Siri, what can you do?"

Dating Siri

If you tell Siri about yourself; for example, where you live, your work address, your family, etc., you can get Personalized service to say things like "Send a message to my wife" and "FaceTime my boss."

Tell Siri about yourself

- Open the Contacts app , then fill out your contact information (*see* Adding your contact info on iPhone (page 136)).
- Go to **Settings** > **Siri & Search** > **My Information**, then tap your name.

Teach Siri how to say your name

Say something like, "Hey Siri, learn to pronounce my name."

Tell Siri about your relationships

Say something like "Hey Siri, Eve Kamin is my wife" or "Hey Siri, Matt Jordan is my boss."

Adding Siri Shortcuts on iPhone

Some apps learn what you frequently do and provide shortcuts for them, so you can ask Siri to do them for you. For example, a workout app may let you view your next appointment with your trainer just by asking Siri, "When is my next training session?"

Adding suggested shortcuts

To add shortcuts to Siri, tap **Add to Siri** when you see a suggestion for a shortcut, then follow the prompts to record a phrase of your choice that performs the shortcut.

Using shortcuts

Say, Hey Siri, then speak your phrase for the shortcut.

N.B. Base on how you use your apps and your routines, Siri also suggest shortcuts on the Home Screen, Lock Screen, and when you initiate a search.

Siri Suggestions

Siri tries to be a step ahead of you by suggesting what you might want to do next, such as confirm an appointment or call into a meeting, based on how you use your apps and your routines.

A Siri suggestion

These are a few examples:

- **Receive calls**: Siri lets you know who might be calling if you get an incoming call from an unknown number, based on phone numbers included in your emails.

- **Start a search or glance at the Lock Screen**: Siri learns your routines so that you can get suggestions for what you need at just the right time.

- **Read News stories**: Siri learns the topics you are interested in and suggest them in News.

Adjusting Siri settings on iPhone

You can adjust or change the default Siri settings to fit your preferences. You can change the voice for Siri, prevent access to Siri when your iPhone is locked, etc.

Changing how Siri responds

Go to **Settings** ⚙ > **Siri & Search,** then do any of the following:

- **Changing Siri's voice**: Tap **Siri Voice**; you can choose a male or female voice for Siri or change the accent (not available in all languages).

- **Changing when Siri provides voice responses**: Tap **Siri Responses**, then below **Spoken Responses**, choose your preferred option.

- **See your voice queries and request onscreen**: Tap **Siri Responses**, then turn on **Always Show Speech**.

- **Always see the response from Siri onscreen**: Tap Siri Responses, then turn on **Always Show Siri Captions**.

Changing when Siri responds

Go to **Settings** ⚙ > **Siri & Search,** then do any of the following:

- **Prevent Siri from responding to the "Hey Siri" voice command**: Turn off **Listen for "Hey Siri."**

- **Prevent Siri from responding to the side or Home button**: On an iPhone with Face ID, turn off **Press Side Button for Siri**. While on an iPhone with the Home button, turn off **Press Home for Siri**.

- **Prevent access to Siri when iPhone is locked**: Turn off **Allow Siri When Locked**.

- **Change the language Siri responds to**: Tap **Language**, then select a new language.

Hiding apps when you summon Siri

To prevent active apps from being visible behind Siri, go to **Settings** ⚙ > **Accessibility** > **Siri**, then turn off **Show Apps Behind Siri**.

Changing where Siri suggestions appear

Go to **Settings** ⚙ > **Siri & Search,** then do any of the following:

- **Suggestions while Searching**
- **Suggestions on Lock Screen**
- **Suggestions on Home Screen**
- **Suggestions when Sharing**

Changing Siri settings for a specific app

Go to **Settings** ⚙ > **Siri & Search,** then scroll down and select an app.

Adjusting the Siri Voice volume

Use the volume buttons or say something like, "Turn up the volume" or "Turn down the volume."

13: iPhone Accessories

The accessories listed below are essential for the proper usage of your iPhone device and are sold separately. iPhone power adapters |iPhone charging cable |iPhone MagSafe chargers |

iPhone power adapters

You can connect your iPhone to a power outlet using its power adapter (sold separately) and a compatible charging cable.

You can charge your iPhone with the following Apple USB adapters. Depending on the country or region, the style and size may vary.

Apple 20W USB-C power adapter

N.B. The iPhone 12 models require a power adapter with a minimum power output of 20 watts for fast charging. If you opt to use a third-party power adapter, it should meet these recommended specifications:

Minimum Power Output: 20 W

Frequency: 50 to 60 Hz, single phase

Line Voltage: 100 to 240 VAC

Output Voltage/Current: 9 VDC/2.2 A

Output Port: USB-C

Apple 18W USB-C power adapter

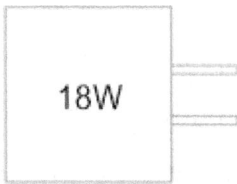

Apple 5W USB-C power adapter

iPhone charging cable

Your iPhone comes with either one of the following charging cables:

USB-C to Lightning Cable

— Ideal for
iPhone 12 models

Lightning to USB Cable

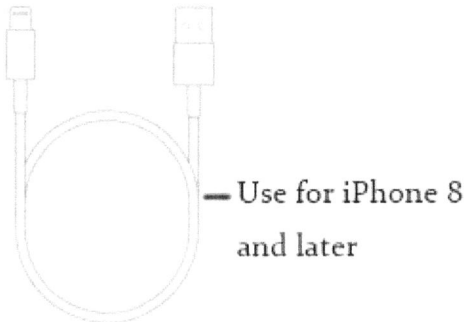

— Use for iPhone 8
and later

Using iPhone MagSafe Chargers

With a ring of magnets encircling the wireless charging coil on the iPhone 12 models, MagSafe Charger (sold separately) magnetically snaps right in place at the back of your iPhone or to its MagSafe case or sleeve (sold separately). Thus, making it possible for you to hold and use the iPhone while charging.

Charging an iPhone or AirPods with MagSafe Charger

You can also use MagSafe Charger with other iPhone models that support wireless charging (iPhone 8 Plus, iPhone X, and later) and with AirPods.

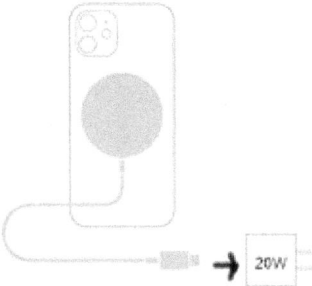

1. Use the Apple 20W USB-C power adapter or any other compatible adapter (sold separately) to connect the MagSafe Charger to power.

2. Do one of the following:

    *   **iPhone 12 models**: Place the MagSafe Charger on the back of the iPhone or its MagSafe case or sleeve. The green charging symbol appears when the iPhone starts charging.

        N.B. Remove iPhone Leather Wallet if it is attached before placing MagSafe Charger on the back of the case.

    *   **Other iPhone models**: Place your iPhone face up on the MagSafe Charger. When you

Align your iPhone properly with the MagSafe Charger, the charging icon ⚡ is displayed in the status bar.

- **AirPods with Wireless Charging Case or AirPods Pro**: Place your AirPods in the charging case, then close the lid, and place the case on the center of the MagSafe Charger with the status light facing up. The status light turns on for several seconds when the case is aligned correctly with the charger, then turns off while continuing to charge.

Using MagSafe Duo Charger to charge iPhone or Airpods

You can charge your iPhone or AirPods at the same time you charge your Apple Watch with the MagSafe Duo Charger. (AirPods, Apple Watch, and MagSafe Duo Charger are sold separately)

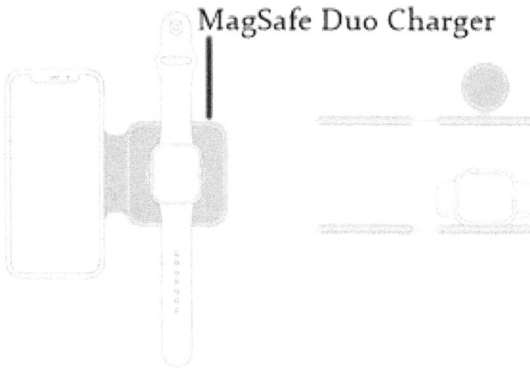

MagSafe Duo Charger

1. Use the Apple 20W USB-C power adapter or any
    other compatible adapter (sold separately) to connect
    MagSafe Duo Charger to power.

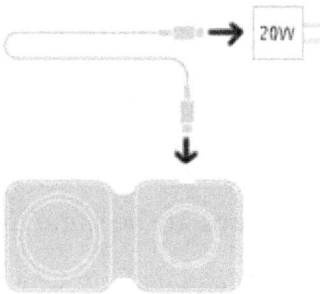

2. Do one of the following to charge iPhone or AirPods:

    • **iPhone**: Place your iPhone face up on the
       MagSafe Duo Charger. On iPhone 12 models,
       magnets help you to align the iPhone with the
       charger easily.

    • **AirPods with Wireless Charging Case or
       AirPods Pro**: Place your AirPods in the
       charging case, then close the lid, and place the

case on the center of the MagSafe Charger with the status light facing up. The status light turns on for several seconds when the case is aligned correctly with the charger, then turns off while continuing to charge.

About the Author

Dylan Blake and Patrick Garner are two Tech enthusiasts or technomaniacs who like to explore the technological world and demystify her mysteries. They have been in this field for more than ten years and are always looking forward to the newest technological advancement. Although they've been friends for a more extended period, their love for tech brought them much closer and made them who they are today.

Made in the USA
Las Vegas, NV
13 March 2022